THE **ARCO** HOW IT WORKS SERIES

LIGHT *WAVE* OF THE FUTURE

Allan Maurer

ARCO PUBLISHING, INC.
NEW YORK

*To Virginia, who made it possible for
me to do what gives my life meaning; to my
mother, Mildred Steinruck, who taught me
to love books and reading; and to Julie and
Rob (Captain Kirk) Mirek, who will live in
the laserlike world of the future.*

Published by Arco Publishing, Inc.
219 Park Avenue South, New York, N.Y. 10003

Library of Congress Cataloging in Publication Data

Maurer, Allan.
  Lasers: light wave of the future.

  (The Arco how-it-works series)
  Includes index.
  1. Lasers—Popular works. I. Title. II. Series:
Arco how it works series.
TA1520.M38   621.36′6   81-7939
ISBN 0-668-05298-8 (Cloth edition)   AACR2

Printed in the United States of America

Contents

Acknowledgments

The writing of any nonfiction work often becomes an almost collaborative effort. Many people, from scientists who granted long interviews, and corporation public relations personnel who arranged interviews and supplied photographs, to personal friends who read the manuscript or provided artwork, were unfailingly helpful. I owe personal thanks to my ex-wife, Virginia, for making it possible for me to write at all; to Renee Wright, for doing library research, transcribing tapes, reading the manuscript, and keeping me from suffocating from lack of human contact; to Betty Wright for doing diagrams, cartoons, and other assignments at the last minute; to Richard Savage for gently whittling away some of my rough edges and showing me whole new worlds; to John Godwin, for humorous cartoon ideas; to Dennis Smirl for photographic services; and undoubtedly to many others.

Many scientists working with lasers were kind enough to take time from their research to talk with me. Dr. Arthur Schawlow of Stanford University was particularly gracious in granting several lengthy interviews and supplying copies of the numerous papers he has written over the years on the past, present and future of laser technology. Other major interviews included those with Mr. Gordon Gould, now at the Opetel Corporation; C.K. Patel and C.V. Schenk of Bell Labs; Dr. John Asmus at Maxwell Labs; and Dr. V.L. Pollack, U.N.C.C. Physics Department. I would like to thank each of them, as well as the many others who clarified points of science, history, law and medicine in briefer discussions.

George Moffett at Bell Labs deserves special thanks for his help in arranging interviews and providing photographs from Bell's files. The public relations departments of IBM, Hughes Aircraft, Raytheon, Spectra Physics and Coherent Radiation all provided vital information and photos. Other illustrations were provided by government sources, including the Department of Energy, the Department of Defense, NASA and DARPA.

Any errors or omissions that remain are the responsibility of the author.

Introduction

It is not often that man creates something totally new. Yet when scientist Dr. Theodore Maiman exposed a bar of synthetic ruby to the flash of a powerful lamp in 1960, the beam of red light that shot across his laboratory was something new. "As far as we know," Dr. Isaac Asimov has written, "it was a variety of light never before seen by the eyes of man . . . a variety of light that never existed on earth before, or in any part of the universe we can see."

It was the first laser beam.

The word *laser* is an acronym formed from the first letters of the words: **L**ight **A**mplification by **S**timulated **E**mission of **R**adiation. Some electrical engineers feel this is a misnomer (wrong name) because lasers actually generate a special kind of light rather than amplify light from an outside source.

What is so special about this light? Unlike ordinary light, a laser beam is coherent. The word *coherent* means "sticks together," and in a sense that's what laser light does. Ordinary light resembles the incoherent babble of voices at a party—it travels in all directions at once. A laser beam, however, is more concentrated than a siren blast, its waves more in harmony than the voices of a trained choir.

By concentrating the energy of its *photons* (light particles) in a narrow, coherent beam that spreads very little, the laser can produce a powerful spike of light capable of sizzling through steel. Focused to a pinpoint spot, it can deliver temperatures three times higher than those at the surface of the sun.

While ordinary white light is a mishmash of all the colors, laser light is *monochromatic* (one-color). This special property of laser light is useful, since all substances absorb the light energy of some colors much more strongly than others. The pure, blue-green light of the argon-ion gas laser, for instance, performs many delicate jobs for surgeons because of the way it is absorbed by certain human tissues; it may prove useful in undersea communications because of the way it is transmitted in water.

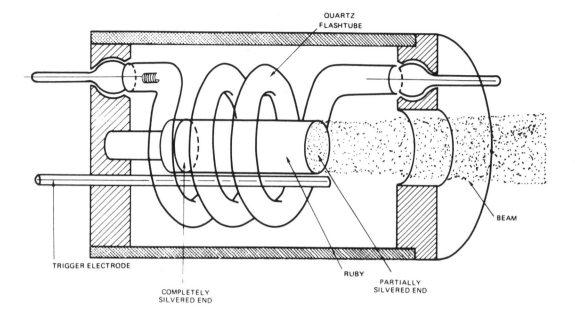

QUARTZ
FLASHTUBE

TRIGGER ELECTRODE

COMPLETELY
SILVERED END

RUBY

PARTIALLY
SILVERED END

BEAM

A ruby laser's basic design isn't complex. A long ruby rod with mirrored ends (one only partially reflecting) is wrapped in a powerful flashlamp. When the flashlamp is tripped, its light excites billions of chromium atoms in the ruby from their normal *ground* (lowest) *state*. When more atoms are excited than at ground state the flashlamp has caused a *population inversion*. As the atoms begin to return to the ground state, they refund the energy from the flashlamp as photons of light in a process called *stimulated emission*. An excited atom struck by one of these photons gives off two photons as it returns to the ground state—one photon refunds the flashlamp's pumping energy and the other is the original photon that struck the excited atom and knocked it back to the ground state. These two photons are not shot from the atom haphazardly: they emerge at the same frequency, locked in step. And each of these photons then strike another excited atom, producing still more *coherent radiation*. Since the chain reaction occurs at the speed of light as photons bounce between the parallel mirrors, a pulse of laser light bursts from the partially reflecting end of the laser in only a fraction of a second. (Credit: NASA)

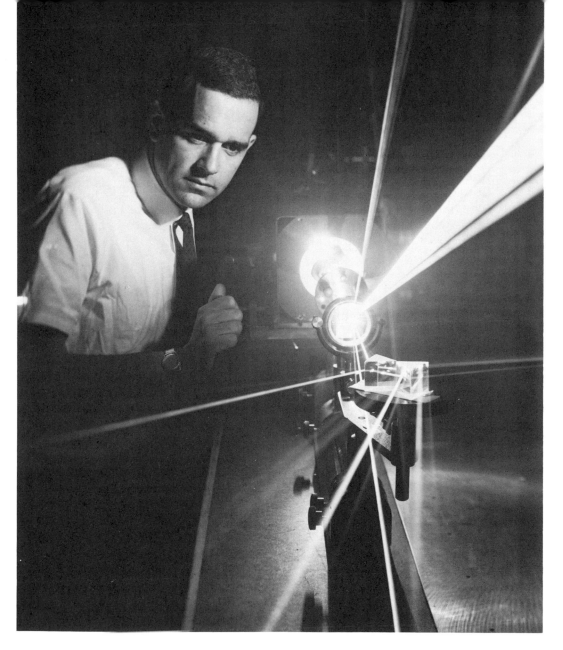

Laser light pours in searchlight-like patterns from a gas laser. This high power, blue-green coherent beam comes from a gas discharge tube located inside the magnet solenoid (large circle in the rear), passes through the output mirror (next circle), is dispersed into several colors by a grating (third circle from rear), and further dispersed by glass prisms in the foreground. (Credit: Hughes Aircraft)

To understand these and other special properties of laser light, it is necessary to know something about the way scientists describe the nature of light, the atom and other natural phenomena. Although mathematics and technical language are required to give a full and precise picture of the way modern physicists view the world of lasers, it is possible to gain a basic understanding of what's happening without such terminology. Keep in mind, however, that the explanations that follow are simplified. The language of physics is much more exact (and exacting) than that used throughout this book.

HOW LIGHT IS EMITTED

We live in a vibrating universe. The atoms of every substance collide, bounce and dance constantly. All of this activity causes the *electrons* of atoms to vibrate (electrons are the tiny charged particles orbiting the heavy central nucleus in the familiar planetary model of the atom). These electrons radiate vibrations of electromagnetic energy. Everything physical in the universe—from the smallest particle to the largest star—radiates waves of electromagnetic energy at a *frequency* (number of up-and-down vibrations per second) depending on temperature.

When the temperature is high enough, radiation is visible as light. Heat the metal filament inside a bulb, the wick of a candle, or the burner on a stove, and light is produced. This occurs because the electrons of atoms move to higher energy states when a substance is heated, then fall back to a normal state as soon as possible, giving back the extra energy as light photons. This is somewhat analogous to stretching a steel spring; the spring will return to normal size rapidly if not held in place.

As an electron of an excited atom returns to its ground state, it gives off a photon of light. These photons, and all electromagnetic energy, differ in the frequency of their vibration and corresponding *wavelength*.

Although trains of waves may be very long (stretching from a star trillions of miles away to earth, for instance) the term *wavelength* refers only to the distance between two crests or two dips as a wave passes a given point. Frequency and wavelength are related: as a frequency increases, wavelength decreases, so that the product of the two is constant—the speed of light in a vacuum, or 186,000 miles a second.

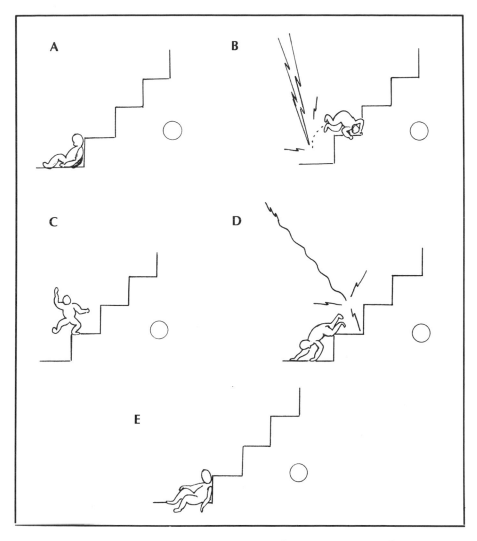

An atom is mostly empty space with a massive central
nucleus (the circle in the diagrams). Electrons travel around
this central core in wavy orbits that go up energy level steps.
Each step is one specific distance above another. (A) An
electron struck by outside energy will be kicked to a higher
step (B). It will teeter there for a tiny fraction of a second (C),
then fall back to the original ground state, giving back the
energy that boosted it up in the first place (D, E). This is
called spontaneous emission. These diagrams do not represent the way an atom really looks. Rather, they are highly
simplified suggestions of the way atoms emit light.

Ordinary white light contains waves of many frequencies and wavelengths that are out of step—like football players after the ball is hiked. Laser light, however, contains waves of a single frequency and wavelength (or nearly so), marching in step like a precision halftime band. Incoherent, ordinary light spreads quickly as its waves travel off in different, ever-widening directions. The waves of laser light, though, travel in phase and in the same direction. They spread so little that when scientists fired a laser at the moon through a telescope, the beam widened to only two miles in the quarter-million-mile journey.

HOW LASERS WORK

How do lasers create this remarkable light? We will deal with this question in more detail in later chapters; however, the basic laser principles are quite simple.

When excited atoms give off photons of light normally, the process is called *spontaneous emission*. But scientists discovered that another phenomenon sometimes occurs: If one excited atom is bumped by another excited atom, two photons are released—one by each atom—and they are locked in step, traveling in the same direction with the same frequency and wavelength. This is called *stimulated emission*.

Although scientists have known about stimulated emission since the 1920s, they did not know how to use the principle to make a laser until the late 1950s, when Dr. Charles Townes and Dr. Arthur Schawlow explained the basic idea of the laser. They suggested that it should be possible to "pump" the atoms in some substances (such as ruby) with a powerful flashlamp to create a *population inversion*—to have many more atoms excited than at the ground state. But the flashlamp alone would not completely do the trick. By adding mirrors at each end of the ruby crystal or laser chamber, the photons released when the flashlamp fired would bounce back and forth in straight lines, stimulating still more excited atoms, to give off coherent photons—thus creating a chain reaction.

After only 20 bumps, a million coherent photons exist, and the pumping flashlamp excites billions of atoms at once. With this process occurring at the speed of light, photons traveling exactly parallel to the mirrors quickly form a pulse of coherent laser light. One mirror is only partially reflecting and permits this beam to emerge.

In the two decades since Maiman created the first crude ruby laser, the device has become one of the most useful tools ever in-

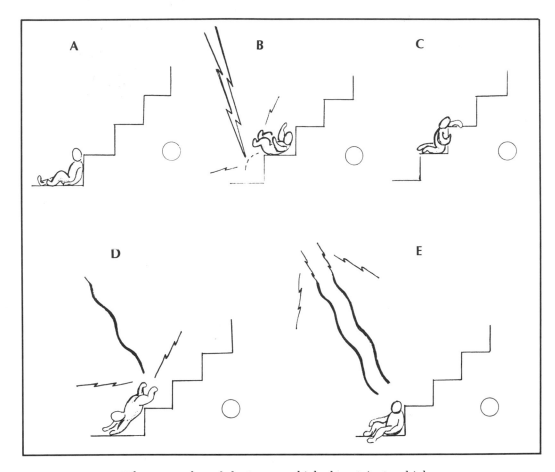

When a number of electrons are kicked upstairs to a higher energy state by the flashlamp or other pumping energy of a laser, the atoms are primed for a chain reaction that will create a beam of coherent light (A through C). A photon spontaneously released by another atom that strikes a still excited atom (D) knocks the electron to the ground state before it has a chance to fall back on its own. This produces the "*stimulated emission*" of two photons of light (E). When this stimulated emission occurs, the photons (one of which refunds the flashlamp energy that originally kicked the electron to a higher orbit, the other being the photon that bumped the electron back to the ground state) come out locked in step. By creating a chain reaction of photons produced by stimulated emission, the energy used to pump a laser is partly changed into a more useful form: coherent laser light.

The Shiva laser, built by Lawrence Livermore Laboratory, is one of a series of high power laser systems built for fusion energy research. The photo shows only six of Shiva's 20 laser amplifier chains. Shiva can deliver more than 30 trillion watts of optical power in less than a billionth of a second to a tiny fusion target the size of a grain of sand. (Credit: US Department of Energy [DOE])

vented. Today they are used to drill diamonds, steel, tunnels and teeth. They vaporize cancerous tumors, repair damaged eyes and provide the blind with "seeing-eye" canes. Lasers weld tiny electronic parts and huge steel plates on ships. They make three-dimensional holographic images, project huge pictures on buildings, print newspapers, slice silicon chips used in calculators and perform a host of other tasks. They have proved useful in virtually every field of human endeavor: art, science, work, play, war.

Today's lasers range in size from microscopically small semiconductor versions (smaller than a grain of sand) to the fusion lasers Shiva and Nova, five times larger than a typical suburban home.

Honest Abe—It's a laser. A semiconductor laser placed on a penny shows why scientists sometimes joke about being careful "not to inhale them." These tiny lasers, which can be made smaller than a grain of salt, are extremely useful in communications work. (Credit: Bell Labs)

The first 20 years of laser research emphasized developing new technology—finding materials that lase, increasing power, creating optics to control the beam. Now the laser is being put to work. In the future it may help create new energy sources, fly airplanes, carry hundreds of radio and TV stations into homes and even launch spaceships.

And, despite its enormous potential as a useful tool, the laser shares with nuclear power—that other two-edged sword of modern technology—an equally large capacity for destruction. Touted by the media as the "death ray" science fiction made famous when laser was first demonstrated, its weapons are now a reality. Yet even as a weapon, the laser may eventually prove to be a boon rather than a curse, by protecting us from nuclear attack.

In the following chapters, we will explore the ways in which lasers affect our lives now and may affect us in the future—from the supermarket counter to the space program.

PAST, PRESENT AND FUTURE

Here, you will encounter the laser story in a variety of ways. I have used words like a time machine, to delve into the past and peek at what may lie ahead in the future. Laser technology and applications are advancing so quickly—especially in the medical, communications and industrial fields—that by the time you read this, some of what is discussed in "future tense" may have already merged with the present. In other cases, always clearly noted, the futuristic scenarios presented are mere suggestions of the possible.

In the relatively new science of futurism, scientists write many speculative scenarios, covering everything from the worst that could happen to the best that is possible—so far as they know or can imagine. Here, admittedly, you will find an emphasis on the positive futures possible. For technology and science (like most of man's tools), whether lasers or nuclear power, are neither good nor evil in and of themselves. It is the way we use them that counts. Personally, I suspect that being aware of the bright, laserlit futures that are possible is an incentive for us to try harder to get there.

THINGS YOU NEED TO KNOW

For your convenience, a few of the terms used throughout this book are explained here for easy reference.

Angstrom (abbreviated Å)—The wavelengths of light are very small (0.00000075 meter or about 1/45,000 of an inch for red light, which has the largest wavelength in the visible spectrum). Therefore, scientists sometimes use a special unit called the *angstrom* to measure these wavelengths. An angstrom is 1/10,000,000,000 of a meter. An atom is roughly 1 angstrom in size, while a typical virus is approximately 100 angstroms. Visible light extends from about 7,500 Å (red) to 4,000 Å (blue). The grooves on a long-playing record are separated by about 100,000 Å.

CO_2 laser—The carbon-dioxide gas laser is the workhorse of the laser field. It emits an invisible infrared beam and is very efficient.

Infrared radiation—Ranging from the edge of the red end of the visible spectrum to 1 millimeter in wavelength, infrared radiation is composed of invisible, long heat waves. On the opposite end, the visible spectrum is bounded by **ultraviolet** radiation—very short waves which are also invisible. Lasers of various types can coherently produce one or the other of these forms of invisible radiation as well as visible beams.

Metric system—Although we have used metric terms sparingly, the following comparisons might prove useful: a centimeter is .3937 inches; a meter is 1.0936 yards; a kilometer is .6214 miles.

Chapter 1

The Sharpest Scalpel: Lasers In Medicine

Born with clusters of growths called papillomas on his vocal cords, two-year-old Robert has not made a sound for more than a year. Viewing the cauliflower-like growths through a microscope in the operating room, Dr. Herbert Dedo targets one with a tiny dot of red light, then steps on a pedal that fires a carbon dioxide laser.

Each split-second pulse vaporizes papilloma tissue in a puff of smoke. The CO_2 laser beam itself is invisible, so Dr. Dedo uses a joystick controller to sweep the red guide light over the growths clogging Robert's throat. As he sees the vocal cords' healthy tissue emerge, "I feel like Michelangelo sculpting," says Dr. Dedo. "When you hear the zap of that laser and think of the precision and power of this light, it seems more like *Star Wars* than anything else."

When Robert awoke in the recovery room following the operation, he cried softly. "It was the most beautiful sound I'd ever heard," said his mother.

Fortunately for Robert and many other persons suffering from conditions that couldn't be satisfactorily treated, it is not necessary to go to a *Star Wars* galaxy "far, far away" to benefit from laser technology. The laser's potential as a science-fiction weapon may receive more media attention, but it is already a ray of life and hope in medicine. One laser inventor, Gordon Gould, quips that "One of the ironies in all this is that the first application of a laser 'death ray' was to mend a detached retina in the eye . . . delicious!"

This first laser application was accomplished at Stanford University in 1963 by doctors Milton Flocks and Christian Zweng. Dr. Arthur Schawlow, co-inventor of the laser concept, also at Stanford, liked to demonstrate one of a laser's characteristics in an amusing experiment. Schawlow aimed a Buck Rogers-styled laser gun at a blue balloon inflated inside a transparent one. When he fired, the ruby beam passed harmlessly through the transparent balloon and exploded the blue one inside. While the experiment amused audiences, it also demonstrated that a laser beam "selects" its target. Because laser light is one pure color (monochromatic) and of a single wavelength, it will be absorbed only by certain colors (depending on the color of the laser beam). In the case of the ruby laser's red light, only the dark blue balloon absorbed its energy.

Flocks and Zweng saw a use for this aspect of laser light in eye surgery. "There's an analogy between the balloons in Schawlow's experiment and the structure of the human eye," Dr. Flocks has explained. Covered with an outer, transparent envelope called the cornea, the eye is backed by the dark-colored, blood-webbed retina. Several eye problems involving the retina were extremely difficult to treat in the early 1960s simply because it's located inside the eye. Retinas become detached from the back of the eye sometimes, a condition that can lead to blindness or severe vision impairment. Ruptured blood vessels in the retina occur frequently in diabetes victims and occasionally from other causes.

After experimenting with monkeys, Flocks and Zweng decided that a ruby laser could be used to spot-weld detached retinas back in place and seal ruptured blood vessels without damaging the cornea. When they aimed their laser through the pupil of a human patient's eye for the first time, the doctors felt a certain tense expectancy. Firing the beam, they saw it pass harmlessly through the patient's cornea and weld the retinal tear. "It took only minutes, and the patient didn't feel a thing," remembers Flocks.

Since that initial use of a laser for medical purposes, it has become one of the most versatile tools in the field since the invention of the x-ray machine. Ruby lasers, too deep a red to be effectively absorbed by human tissue, have been replaced by the powerful carbon dioxide device (*see* How It Works: CO_2 Lasers in Medicine, p. 9) and the argon gas laser and YAG (yttrium, aluminum, garnet crystal) machine.

Both the argon and YAG lasers can be operated in the visible light region to produce a green beam (at a wavelength of five thousand angstroms—an angstrom equals one hundred-millionth of a centimeter). Although these lasers do not produce the high-powered cutting beam that makes the CO_2 laser the "workhorse" of medicine and many other applications, their green light and low power are ideal for eye surgery.

The first time an argon laser was used on a human patient, it presented a nine-year-old girl with a very special Valentine's Day gift. Suffering from a membrane growing across her eye and slowly darkening her world, she had undergone three previous operations

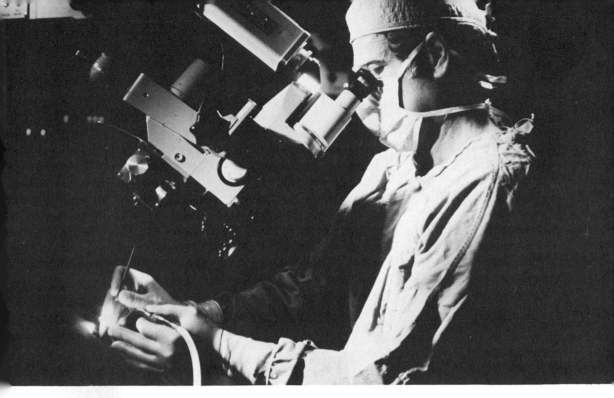

An argon laser developed by Spectra Physics is widely used to perform nearly bloodless surgery. Here a doctor sighting through a special microscope operates on a patient's ear. (Credit: Spectra Physics)

which failed. Using conventional methods, doctors could not continue their attempts to remove the membrane with a scalpel; extensive bleeding forced a stop. In February 1968, Dr. Francis A. L'Esperance of Columbia Presbyterian Hospital cut through most of the membrane with an argon laser scalpel. The laser's heat sealed each blood vessel as he worked, preventing bleeding. The next day, surgeons completed the operation successfully. By February 14, the girl could see again.

Since then, the argon lasers have become a common tool in eye surgery and other medical applications. A typical operation to seal leaking blood vessels in the eye might go like this: doctors dilate the patient's pupil with eyedrops; then, a special contact lens is inserted to magnify the blood vessel-laced retina; finally, argon laser beams projected through the eye perform up to 4,000 spot-welds, each no larger than the diameter of fine human hair. Referred to as "pattern bombing," this procedure is so precise that the burns do not affect the patient's vision.

The argon laser has found numerous other uses in surgery, primarily due to its ability to stop bleeding. Its cool green beam coagulates blood without burning it away. In addition, it can be focused

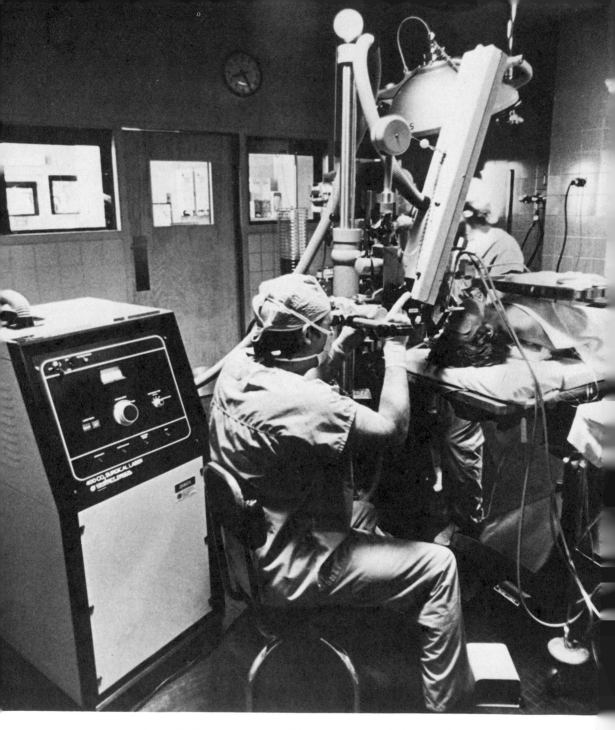

A surgical laser may resemble a dentist's drill, but unlike that often feared instrument, the laser works so quickly and effectively that it usually results in less pain, shorter recuperation time, and fewer complications than the surgeon's steel knife. (Credit: Spectra Physics)

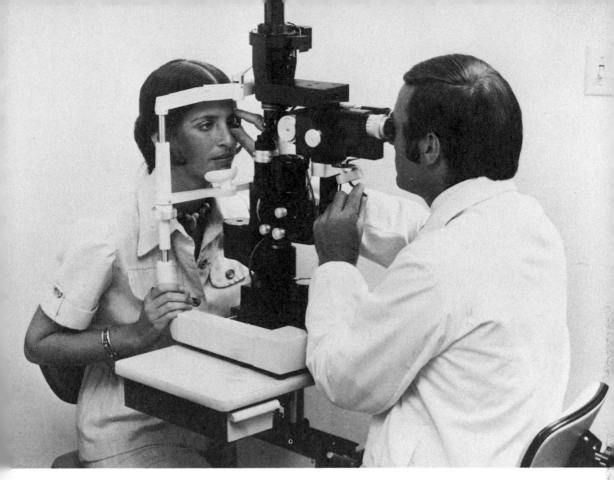

Lasers have proved useful in eye surgery because a laser operation is often so quick and painless that it can be performed in a dispensary or doctor's office rather than an operating room. (Credit: Spectra Physics)

precisely enough to cut through the skin of a tomato without touching the meat. West German scientists took advantage of these features to use argon laser beams to stop potentially deadly internal bleeding in human patients. In the U.S., Dr. Albert M. Waitman, of Beth Israel Hospital in New York, has pioneered this use of lasers for the same purpose.

In a typical operation, a patient suffering from internal bleeding swallows a three-and-a-half foot snakelike tube. Called an endoscope, it contains thousands of fiber-optic threads that reveal a magnified image of the patient's gastrointestinal area to the doctor. Looking through the eyepiece at the top of the tube, the surgeon guides a separate fiber that will carry the laser beam into the area. When the laser is fired, its light is absorbed by the hemorrhaging blood vessels, which are sealed by the resulting heat.

Still another use of the argon laser is being tested in several U.S. hospitals: treatment of "port-wine" birthmarks. Also called strawberry marks, these reddish-purple stains often appear on the face and neck. Caused by visible clusters of blood vessels under the skin, they are medically known as hemangiomas. In the past, doctors treated such blemishes with x-rays, cryosurgical freezing or surgery—all solutions with potentially dangerous side effects.

Ten years ago, however, a California plastic surgeon, Dr. Harvey Lash, suggested that argon lasers might be used to treat these blemishes without side effects as dangerous as those in older methods. A birthmark of the port-wine type scales off like sunburnt skin when the argon beam strikes it. This leaves a white area that may darken to normal skin color. Doctors have lightened blemishes of this sort on two-thirds of the first 100 people treated at the Palo Alto Medical Clinic.

In the treatment, a laser is focused through a special device called a breoptic tube. A patch of skin no larger than two square inches is treated with each blast. Two or three laser treatments will lighten a small stain; as many as a dozen may be necessary to eliminate larger ones. The laser surgery is performed on an out-patient basis under local anesthesia. Although results so far are promising, doctors caution that even this treatment poses the risk of scarring, or skin mottling caused by laser burns.

EAR SURGERY

Doctors at Palo Alto also use argon lasers to operate on the inner ear, a location almost as hard to reach through conventional surgery as the retina. Dr. Rodney Perkins developed the idea for one procedure while watching a television program on lasers. He now uses the device to correct a common hearing defect caused by an abnormal bony growth in the middle ear. The disease, otosclerosis, frequently damages the hearing nerve after many years.

Perkins corrects this defect by operating with the laser to remove the stapes bone from the ear and replaces it with a stainless steel wire. The wire then allows sound vibrations previously blocked by the bony growth to reach the eardrum, and thus the brain. Perkins performed this operation more than 30 times between 1978 and 1980. Patients undergoing the laser surgery experience less bleeding and suffer less dizziness than is common in conventional operations.

Perkins believes laser techniques can be used to treat other defects within the ear as well.

THE LIGHT SCALPEL

Although the low-power, blue-green light of the argon laser is amazing enough, the continuous high power available from the CO_2 laser is having an even larger impact on many surgical techniques. The invisible infrared beam of the CO_2 instrument is entirely absorbed by water, explains its inventor, Bell Laboratories scientist Kumar Patel. Since the body is 75 to 90 percent water, the CO_2 laser can be used nearly everywhere in the body.

The CO_2 laser scalpel offers numerous advantages over the surgeon's steel knife. It can be focused finely enough to vaporize a single cell. It cauterizes blood vessels as it cuts, reducing or eliminating heavy bleeding which can endanger the patient's life and obscure the surgeon's view. It causes no contact pressure during cutting. And, it is fast. These combined features help reduce the shock, trauma and disfigurement of surgery in places where the laser can be used.

Operating with a Rainbow

Prominent laser surgeon Dr. Geza Jako has stated that current technology has allowed doctors to explore no more than 20 percent of the possible medical applications of lasers. Reasons for this include the high cost of laser systems (from $15,000 to $90,000).

At Sinai Hospital, however, Drs. Hugh Beckman and Terry Fuller have engineered a laser operating room system that may help lower these high costs and increase the use of lasers in surgery. Their system locates all of a hospital's laser units in a single room. From there, fiber-optic cables pipe laser beams to different operating rooms, where doctors can guide them. This eliminates the need for complete laser units and technical staffers in every operating room and greatly reduces the cost of making the instruments available to surgeons.

It works like this: The surgeon plugs a fiber-optic cable into the operating room wall panel and selects the laser desired. With this system a doctor may even switch from one type of laser to another, as necessary, during an operation. The surgeon then steps on a control pedal to activate the laser, which instantaneously shoots through the threadlike fibers to do his bidding. He has a rainbow at his fingertips.

The speed and precision of the CO_2 laser are particularly beneficial in two medical fields where delicacy and quickness are vitally important: neurosurgery (brain surgery) and gynecology (the treatment of women's diseases).

Dr. Joseph H. Bellina, professor of obstetrics and gynecology at Louisiana State University and director of the Laser Research Foundation, has pioneered the use of lasers in gynecology. After his first use of laser surgery as a last resort in 1974, Dr. Bellina was "astonished" at the way his patients recovered following their operations. Operating on a woman's reproductive organs (for example, the cervix, vagina and vulva) requires keeping tissue damage from inflammation and scarring to a minimum. Otherwise, a woman may become sterile—unable to have children. Because the laser scalpel seals tiny capillary blood vessels as it cuts, it causes less swelling, pain and scar formation, while leading to faster healing. "This is a major advance," asserts Dr. Bellina.

His early work with laser microsurgery proved so successful compared to conventional methods—9 out of the first 12 patients on whom he used the techniques later had children—that he formed the Laser Research Foundation in 1979. He has since used the laser scalpel in more than 1,400 operations, originating many procedures himself. These include operations to remove both benign and cancerous tumors, although Dr. Bellina firmly points out that here even the laser has serious limitations.

Perhaps the most impressive feat Dr. Bellina—or any other surgeon—has performed with a laser is his reconstruction of a woman's reproductive anatomy with a knife of light. That's a sculpting job even Michelangelo would have been pleased with. Whether dramatic or commonplace, however, most of the operations Dr. Bellina has performed or supervised "would not have been done several years ago," according to him. "They would have been surgical rejects because of the trauma and operating time needed. You just can't keep a patient under a general anesthetic that long for an elective procedure."

Laser surgery cuts operating time by up to 50 percent by eliminating much of the need for clamps, sponges and scalpel incisions with resulting bleeding problems. Many of Dr. Bellina's operations are for "fertility enhancement," to help women who cannot have children for anatomical reasons. He did 61 such operations with a laser in 1980 alone, with good success. "We can now offer hope to hundreds of women who would otherwise go childless," he said with a justified sense of pride.

How It Works: The CO_2 Laser in Medicine

Developed at Bell Labs in 1964, the carbon dioxide (CO_2) laser does the heavy work where high power output is required—in medicine, as in many other areas. "When you talk about applications in laser surgery," says C.K.N. Patel, inventor of the device, "where light can produce sufficient heat to cut a tissue, a key factor is the high power of the CO_2 beam."

The CO_2 laser operates in the infrared wavelengths producing an invisible ray. A red guide light directs the surgeon's hand when it is used as a scalpel.

Since water absorbs infrared light almost totally and the human body is 75 to 90 percent water, the CO_2 laser can be used on nearly all human tissues. The laser scalpel cuts by heating the water in cells extremely rapidly, turning it into steam. As it expands, the steam explodes the cell. Any tiny stray pieces are vaporized or burned, a factor useful in removing cancerous tumors. Also, the laser beam seals small capillary blood vessels as it cuts, by cauterizing them. This is one of the laser's most helpful aspects, for it stops massive, life-threatening bleeding, saves time, reduces infection and speeds healing.

Although a surgeon wielding a steel scalpel can cut as quickly as one using a laser knife, time in the operating room is reduced with the light scalpel. Less mopping up of blood, clamping of leaking blood vessels and so on are needed in laser surgery.

Mounted in the operating room, a CO_2 laser vaguely resembles a dentist's drill hanging from its mobile chassis. Surgeons aim the laser's beam in several ways. A hand-held penlight-sized device can be used much as an ordinary scalpel would be. When surgery requires the use of special microscopes or other viewing instruments, the laser knife can be controlled with a joystick and foot pedal. Accurate enough to blast a single cell without harming surrounding tissue, the laser can be focused to a depth or width of one millimeter (about four hundredths of an inch—try marking that off on a ruler).

ZAPPING BRAIN TUMORS

At the State University of Buffalo in New York, Dr. Patrick J. Kelly zapped a woman's brain tumor with laser light, vaporizing it. Computerized brain scans after the operation showed no sign of the tumor. In a recent article in the *Journal of the American Medical Association*, Kelly reported that the woman, who is in her fifties, was still in good

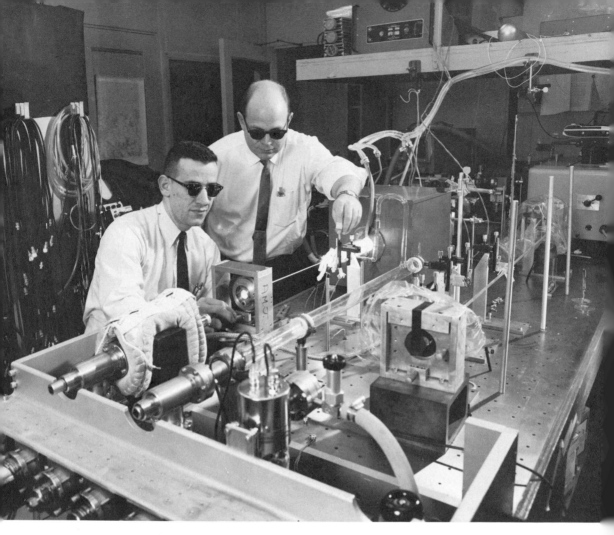

Surgical argon lasers were developed by Dr. E.I. Gordon and
E.F. Labuda at Bell Telephone Laboratories. In this photo,
Labuda and another Bell scientist, A.M. Johnson, are shown
experimenting with the device. (Credit: Bell Labs)

health. Kelly said his technique allows surgeons to destroy a deep-
seated tumor without spreading cancer cells. The laser knife also does
little damage to healthy tissue. Bleeding, which can be very dangerous in conventional brain surgery, was not a major problem because of
the laser's automatic sealing of vessels.

Similar work with lasers has been done at the University of
Pittsburgh, where researchers also removed cancerous growths from
brain tissue. Other doctors have successfully treated victims of Par-
kinson's disease (which affects the nerve center at the base of the
brain) with lasers.

Despite the success many surgeons have had in removing early, preinvasive cancers (in which malignant cells have not spread), Dr. Bellina cautions that they are ineffective against more advanced cancers. He points out that lasers are a "wonderful new medical instrument . . . not a wonder drug."

Dr. Dedo, however, who performed the surgery described at the beginning of this chapter, asserts that "the laser approach to surgery is still in its infancy. We can't even estimate what its potential will be."

PIERCING THE EAR

One new use of the laser, now being tested, is as a treatment for recurring ear infections in children. Richard Goode, associate professor of surgery at Stanford University, believes the laser may slice the cost of treatment in half. Such infections are common among children and some adults. Fluid collects in the middle ear, providing a sticky swamp that bacteria love. The resulting infection can be extremely painful and impair hearing.

At present, the condition is treated by poking a hole in the patient's eardrum, inserting a tube and draining the fluid. This requires sticking the ear with three to six needles to anesthetize it—a procedure most children, especially very young ones, can't stand. Most, therefore, are put under general anesthesia, which requires a stay at the hospital and is expensive.

With the laser technique Goode has tested on animals and several human patients, the eardrum is numbed without needles. Anesthetic is placed in the ear canal and forced to the eardrum by an electric current. Then, a CO_2 laser is aimed with a red guide light and fired. It punctures the eardrum in a tenth of a second. Patients hear a "pop" as the laser beam vaporizes a tiny hole one or two millimeters in diameter. The operation is painless, and the eardrum heals in about a month. Although Goode is optimistic that the laser surgery may benefit as many as half the patients who suffer from these infections, there is some concern that a month may not be long enough for the ear to completely drain.

LIGHT OF THE FUTURE

Nearly every doctor who has worked with lasers notes that many possible medical uses of this helpful, man-made light are still un-

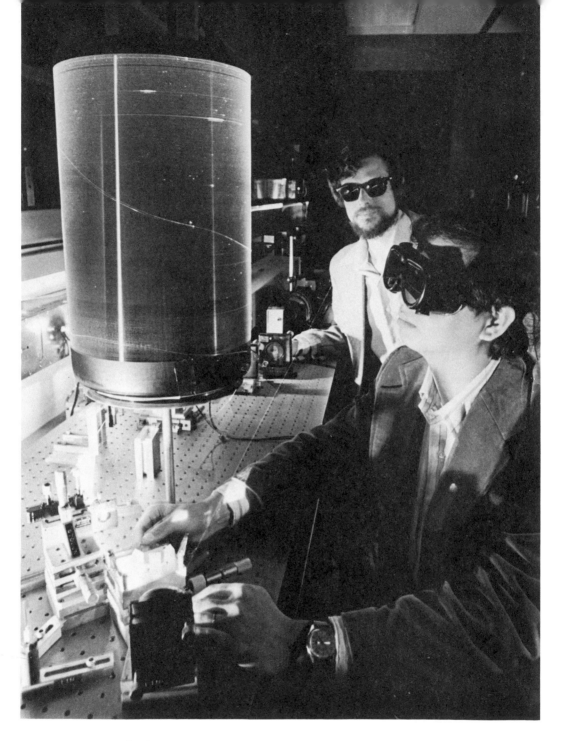

Chinlon Lin and Roger Stolen, scientists at Bell Laboratories in Holmdel, N.J., are shown here aligning the world's longest laser, a one-kilometer optical fiber. It's called the "fiber Raman laser." (Credit: Bell Labs)

tapped. Research continues, and new findings are reported almost daily.

Laser diagnosis of various forms of cancer is one area that looks promising. Many cancers, particularly lung malignancies, must be detected very early or they are often virtually incurable. Each year, 117,000 new cases are found and 101,300 people die from the disease. By the time lung cancer shows up on an x-ray, it is usually so far advanced that it is a death sentence.

Now, researchers at the University of Southern California's Medical Imaging Science Group are testing a new technique with lasers which may be able to detect lung cancers before they are one millimeter long and only 100 millimeters thick. No conventional method of detection can do this.

"Finding hidden or x-ray-invisible cancer is the problem," says Dr. Oscar Balchum. "People who feel good don't go to the doctor."

In the new detection procedure using laser, the patient swallows a bronchoscope (an instrument used to see inside the lungs), which is attached to a fiber-optic cable. The cable transmits the intense laser light to the lungs. Hematoporphyrin-derived drugs, given to the patient earlier, tend to collect in cancer cells and make them appear slightly reddish when hit by the laser beam.

In one test of this technique in Canada, a researcher discovered an almost invisible cancer in the lungs of a 70-year-old farmer. No more than 1.5 millimeters thick, it was removed, and the patient is expected to remain free of the disease.

Other laser imaging systems are being used to identify early breast cancer, which now strikes one out of every four women in the U.S., without the dangers of x-rays. They are also put to work exploring the structure of the human eye and may, some experts believe, provide a new understanding of how human vision works.

The list of medical uses for this almost unbelievably versatile tool goes on and on. At the University of Washington, engineers constructed a laser-assisted sapphire scalpel that reduces blood loss when doctors must remove severely damaged tissue from burn victims.

Doctors at Sinai Hospital are attempting to correct nearsightedness and astigmatism by reshaping the lens of the eye with carefully placed burns on the front of the eye. Because the laser works so quickly, only a tiny area is scorched, painlessly. When the burned area heals, it shrinks, tightening the focus of the eye's previously misshapen lens.

Yet another frontier application of lasers—still in the experimental stage—is their use to slice away the built-up material that clogs arteries, causing heart attacks, strokes, blindness and senility. Here, the laser's ability to select its target and cut to a precisely calibrated depth may be uniquely effective.

Dental researchers are experimenting with laser light in attempts to make tooth surfaces stronger. Lasers pump so much energy at a given area, they can create photochemical changes in many substances. Perhaps in the future, you will lean back in the dentist's chair after having your teeth cleaned by ultrasound and have the enamel toughened by a laser light bath.

With its ability to focus energies rivaling those on the surface of the sun to a point so minuscule it will drill holes in a single human cell, the laser may prove to be a necessary tool in the coming biological revolution. Genetic microsurgery on the very stuff of life, performed by exquisitely controlled laser beams, may help doctors correct birth defects, clone human cells and achieve other genetic engineering marvels. Not only the laser's quick, sharp drilling-and-cutting prowess, but also the effects of its concentrated light on human tissues, may be important in this field. Just as the laser has become an effective medical tool, so too it may become a powerful biological probe; what is useful in one area frequently has applications in the other.

Conferences on the medical use of lasers produce hundreds of abstracts each year on new techniques, approaches and research projects. It is likely that every hospital in the United States, and probably much of the rest of the industrialized world, will soon regard lasers as every bit as necessary as x-ray machines.

Chapter 2

Light on the Past: Inventing the Future

The men and women who invent the future require light from the past. Dr. Theodore Maiman of Hughes Aircraft Company created the first pulse of man-made laser light in July 1960; but in a press conference he said that his achievement "marks the culmination . . . of efforts by teams of scientists in many of the world's leading laboratories."

It should be added that those efforts succeeded because the orderly process of science, from man's first attempts to understand light and matter to Einstein and beyond, mapped the way. In his *Intelligent Man's Guide to Science,* Dr. Isaac Asimov notes that "The scholars of ancient and medieval times were completely in the dark as to the nature of light." Yet men dreamed.

Dr. Arthur Schawlow, one of the co-inventors of the laser concept, has written: "In some ways, lasers seem to be the realization of one of mankind's oldest dreams of technological power. Starting with the burning glass, which was known to the ancient Greeks, it was natural to imagine an all-destroying ray of overpoweringly intense light."

First steps toward a real understanding of the nature of light were taken by Isaac Newton around 1666. He observed sunlight through a prism and discovered that ordinary white light is really a mixture of all the other colors of light. Beamed through two faces of a triangular prism, the white light spreads into a band of colors. This band of light, partially seen in a rainbow when atmospheric moisture bends sunlight into some of its components, is called a *spectrum* (from the Latin word for "ghost," which was itself derived from the Latin *specere*: to perceive with the eyes, to see). The visible light spectrum consists of the colors red, orange, yellow, green, blue and violet.

Newton's experiments convinced him that light was composed of tiny particles he called *corpuscles*. He theorized that white light consisted of a "soup" of these particles, with their different masses and speeds responsible for various colors. His theory explained why light travels in a straight line, bounces off shiny surfaces and outlines sharp shadows. But it left other questions unanswered, such as why two beams of light could cross without interfering with each other. Shouldn't the particles collide, changing the brightness of the beams?

So, in 1678, a Dutch physicist, Christian Huyghens, offered a competing theory. He suggested that light consists of tiny waves rather than the collection of colored bullets Newton proposed. Toss two stones close to each other in a pond and you will see that waves can go through each other without losing their separate identities. Both wave patterns will continue unimpeded. The wave theory also explained why light refracted into different colors when beamed through a prism—harder to explain in the particle theory. If light consisted of waves, the wavelength of each wave would determine how much it refracted. Shorter waves would bend more at the surface of the prism. Thus, waves of violet and blue, Huyghens suggested, have shorter wavelengths than those of red and yellow light. Unanswered questions remained, however. If light is composed of waves, asked critics, why doesn't it go around objects the way water and sound waves do?

Scientists argued these points for nearly a century before further progress was made. Then, in the early 1800s, two physicists made discoveries that seemed to answer those questions. First, Thomas Young, an English doctor who also studied physics, showed that light did, indeed, behave like waves. Young projected a thin spear of light through two close-spaced, narrow slits in a mask onto a screen behind this mask. Light bullets, if they existed, should produce two bright lines on the screen, one for the part of the beam coming through each slit. Light waves, on the other hand, would create bright bands where the two beams coming through the slits were in phase and reinforcing each other. Where the trough, or dip, of one wave met the peak of another, though, they would interfere with each other, producing alternating bright and dark bands. That is what happened: Young's experiment produced exactly such a series of bands on the projection screen, confirming that light behaves like waves.

The experiment not only revealed that light behaves like waves in this respect, but also provided a means to measure the wavelength of light. By measuring both the distance between the two slits and the spacing of the bands which result when light is passed through them, wavelength is determined. This calculation shows that the wavelengths of visible light are very small compared to ordinary objects. This is why such objects cast sharp shadows; even microscopic bac-

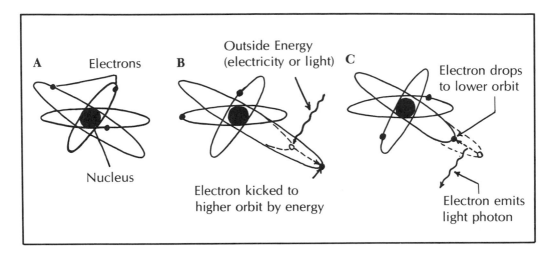

A Electrons   B Outside Energy
                  (electricity or light)   C
                                           Electron drops
                                           to lower orbit

Nucleus

Electron kicked to
higher orbit by energy

Electron emits
light photon

The Danish physicist, Niels Bohr, combined the theories of
Ernest Rutherford, Max Planck, and Albert Einstein to
create his planetary model of the atom. Rutherford demonstrated through experiment that the atom is mostly empty
space with its mass concentrated in its center (the nucleus
in A). Planck and Einstein established quantum theory by
showing that an atom will absorb and give off energy in
certain whole number (quantum) amounts, but never in
fractions of these quantities. In other words, an atom may
absorb a nickel's worth of energy and return a nickel's
worth of energy, or a penny's worth, but never a half-
nickel's worth, or a half-penny's worth. Energy states in
atoms, like monetary systems, are *quantized*. Bohr
suggested that this can be explained by imagining the atom
as something like a miniature planetary system, with the
massive nucleus in the center and the planet-like electrons
orbiting around it. The farther away from the nucleus an
electron orbits, the higher its energy state, Bohr said. Outside energy striking the atom would boost an electron to a
higher energy state orbit (B). Like a piece of wood blasted
into the air by an explosion, the electron wants to return to
its original "ground state" orbit. When it falls, it returns the
energy that kicked it into a higher orbit as a photon of light
(C). Although this planetary model of the atom was later
refined to include the findings of modern physicists, it
explained a great deal about how physical and chemical
reactions happen. It is one of the most useful ideas of modern science, despite the fact that Bohr himself called it only
a "crude beginning" in the understanding of atomic structure. (Credit: Betty L. Wright)

teria are large compared to the wavelength of visible light.

Waves swing around objects only if the wavelength is roughly comparable in size to the object; for example, a large water wave easily moves past a rock, ship or other object that is small compared to its wavelength.

After Young's experiment, the French scientist Augustin Jean Fresnel demonstrated that even if an object is very tiny, light waves will travel around it. Fresnel cut minuscule nicks in a pane of glass, then shone a light through it at a screen. Not much larger than the wavelength of the light beamed through them, the nicks produced blurred and fuzzy shadows. Since waves diffract (flowing around objects rather than sharply defining them) this seemed to settle the question of the nature of light.

Appearances, even in science, can be deceiving. As the 19th century progressed, physicists, mathematicians and inventors made rapid progress. Michael Faraday related electricity to magnetism. James Clerk Maxwell, a brilliant Scottish mathematician, predicted the existence of a vast electromagnetic spectrum of many frequencies of radiation, all traveling at the speed of light (186,000 miles per second). Heinrich Hertz discovered and learned to control radio waves, proving Maxwell correct and fathering the modern communications revolution (for more on all of these events, see Chapter 6, Listen to the Light). Nevertheless, despite these theoretical and technical advances, by the turn of the century scientists were once again questioning the basic nature of this now very useful electromagnetic spectrum.

One question in particular continued to bother scientists: Since all other known waves consist of disturbances traveling through a medium such as water or air, how did electromagnetic energy, including light, travel in a vacuum? If electromagnetic energy is transmitted by waves, shouldn't this radiation have to travel *through* something? How could a wave move through the emptiness of space? Since no one could think of a way for this to occur, scientists proposed the existence of an "ether", a virtually magical substance that was at one and the same time a solid and yet thinner than gas. (Not the chemical that doctors once used to anesthetize patients for surgery.)

MODERN PHYSICS: GOODBYE, ETHER

Modern physics and the discoveries that led to the development of atomic power, the hydrogen bomb, the transistor and the laser began with an experiment that didn't work but did put an end to the ether theory. In 1887, Albert Michelson and Edward Morley performed an experiment to measure how much the so-called ether deflected a light beam. Michelson invented a special device he called an interferometer to do the experiment. The interferometer uses a half-mirror to split a light beam aimed at two other reflecting mirrors. One part of the split beam goes forward while the other veers off at a right angle. When the reflecting mirrors shoot the beams back to the instrument, if one travels even the slightest amount farther than the other, the interferometer reads an interference pattern caused by the out-of-phase waves. This gives a measurement of length so precise it can tell scientists how much a plant grows in a second. When Michelson and Morley performed their famed experiment, they found that there was no evidence that ether existed. In 1907, Michelson became the first American to win a Nobel Prize (physics).

ROAD TO THE LASER: QUANTUM MECHANICS

This presented scientists with the need to re-evaluate their ideas about light. If light and other forms of electromagnetic radiation travel in waves, but the waves are not made through a substance, how are they made? In other words, we know they're waves, but of what?

In 1900, the German physicist Max Planck came up with an answer that led directly to the laser . . . and many other wonders of modern science. Studying what was known as the "black-body radiation problem," Planck discovered that to fully explain the way matter emits radiation when it is heated, he had to return to the idea that electromagnetic particles exist. The problem, essentially, was this: when heated, matter radiates energy. As the matter gets hotter, the energy moves toward shorter wavelengths (from invisible infrared, to dim red, to cheery bright red, to orange and on to yellow-white). But the distribution of energy according to wavelengths changed in unexpected ways that no theories of the time would explain. Although the total amount of radiant energy given off increases as the temperature climbs, peak levels decrease in intensity. No matter how they tried, scientists could not make the math come out right using the wave theory of electromagnetic radiation to explain this.

Planck departed completely from contemporary thought. To mathematically describe the way matter actually emits radiant energy, Planck found it necessary to regard each molecule as a generator that can absorb and transmit only discrete (distinct) packets of energy. Radiation, he said, is made up of these energy packets.

Planck named these energy bullets *quanta* (from the Latin word for "how much"). Each quantum carries a specific amount of energy related to its wavelength; the shorter the wavelength, the greater the energy it carries. The quanta are also directly related to the frequency of radiation; the higher the frequency, the more energetic the quanta. Moreover, Planck said, these quanta can only be emitted or absorbed in whole numbers.

He summed up these ideas with the formula $E = hf$. The E represents energy, the f, frequency, and the h, what is known as *Planck's constant*. This number is so small compared to the very high speed of light—$h = 0.000\ 000\ 000\ 000\ 000\ 000\ 000\ 000\ 006\ 624$—that each quantum packet is very weak, and so tiny that light, like the atoms of solid matter, appears continuous to our eyes.

ENTER EINSTEIN

Planck's quantum theory was too far removed from conventional ideas to take effect immediately. Five years after Planck first outlined his theory, however, Albert Einstein confirmed the existence of quanta. He explained the then recently discovered "photoelectric effect" perfectly by way of the quantum theory.

What is the photoelectric effect? When light strikes certain metals, it knocks electrons (very small, charged particles that orbit the nucleus [center] of atoms) from the surface. At first, scientists studying this effect were puzzled because increasing the intensity of the light used did not add energy to the electrons kicked away, but changing its frequency did. Blue light sent the electrons speeding away faster than yellow light. Yellow light kicked them harder than red, which on some metals had no effect at all. In other words, the higher the frequency of the light, the more energy it transmitted.

Einstein, who received his 1921 Nobel Prize for explaining this photoelectric effect, not for his theory of relativity, renamed light particles *photons* (from the Greek word for "light"). Photons managed to confuse scientists for so long because they are, Einstein suggested, two-faced. They sometimes act as a particle, sometimes as a wave, depending on the circumstances. Science writer Ben Patrusky describes them as "a bullet with a wiggle." In one second, a 40-watt bulb, commonly used for soft household lighting, gives off nearly a quintillion (1,000,000,000,000,000,000) photons, varying in frequency and wavelength.

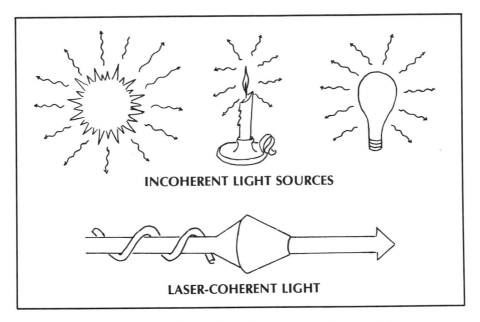

INCOHERENT LIGHT SOURCES

LASER-COHERENT LIGHT

Ordinary light, whether from the sun, a candle, or a light bulb, is incoherent. It travels in many different directions and spreads to invisibility after going a short distance, unless powered by a star-sized nuclear furnace. Point a flashlight into the darkness outdoors and watch the beam spread until it disappears. Albert Einstein and Paul Dirac pointed out that a different kind of light is produced when a photon of the proper frequency strikes an excited atom. By causing a chain reaction of photons produced by stimulated emission, a laser sends out a beam of coherent light in a straighter, truer line than any shaft Robin Hood ever loosed. (Credit: Betty L. Wright)

Perhaps they are best pictured by imagining a light beam composed of many wavy bits (as opposed to a single wavy line). Each of these wavy light bullets has a *length* (distance from one wave peak to another) and *frequency* (number of wave peaks which pass a given point in one second). Short wavelength photons have more peaks and more energy packed into their wavy bullets than longer ones.

All forms of electromagnetic radiation have these characteristics. They are arranged according to decreasing wavelength (*see* chart, page 82) on the electromagnetic spectrum. They range from the ultra-low-frequency radio waves thousands of kilometers long (a kilometer is about half a mile) to the incredibly tiny and energetic cosmic rays. In between are broadcast radio waves measured in hundreds of meters (a meter is slightly more than a yard); radar microwaves measured in centimeters (a ten-centimeter radar wave is about four inches long); the light we can see, measured in angstrom units (an angstrom unit, abbreviated Å, equals one hundred-millionth of a centimeter); and the still shorter, invisible ultraviolet, x-ray and gamma-ray bands.

The visible part of the spectrum extends from 7,500 angstroms (red) to 4,000 Å (blue). If the entire spectrum thus far observed were represented by a 12-inch ruler, the visible portion would take up only one-sixteenth of an inch. Since the energy delivered by a given quanta increases as wavelength decreases, it can be seen that photons of red light are much less energetic than those of blue. Only half as energetic as those of violet light, red photons are not powerful enough to expose photographic film by activating its chemicals. That's why red "safe" lights can be used in photographic dark rooms. Higher-frequency, shorter-wavelength ultraviolet light, invisible to the naked eye, is so strong that it can cause sunburn or damage to the eye if one looks directly at the sun.

Sunlight, by the way, contains a broad range of electromagnetic radiation, from radio waves through the entire visible band to the ultraviolet. Any light containing photons of many different frequencies and wavelengths is called *incoherent light*. While not containing as large a variety of frequencies as the sun, ordinary electric light, candlelight and moonlight are all soupy collections of photons that vary in frequency and wavelength; thus they are incoherent. This is one of the major ways laser light differs from ordinary light . . . but we'll come to that later.

PROBING THE HEART OF THE ATOM

The discoveries of Planck and Einstein led other scientists to construct models of nature's building block, the atom, to explain how it could emit particle wave photons. By 1916, Ernest Rutherford, an English physicist, and Niels Bohr of Denmark, had pictured the atom as a heavy central nucleus orbited by electrons traveling in fixed paths around it.

In Bohr's model of the atom, an electron can orbit in any of the given fixed paths, but never between them. Each of these fixed shells in which the electrons can orbit represents an energy level. Electron energy levels increase in shells farther from the nucleus, and decrease in shells closer to the nucleus. Boosted by outside energy, heat or light for instance, an electron can jump from an inner shell to an outer one at a higher energy level, but never to an in-between position. This explains why quanta are emitted in certain whole-number sized packages only.

Atoms with electrons lifted to higher energy level shells by outside energy are described as being in "an excited state." With all of its electrons orbiting at their lowest levels, the atom is in the "ground state."

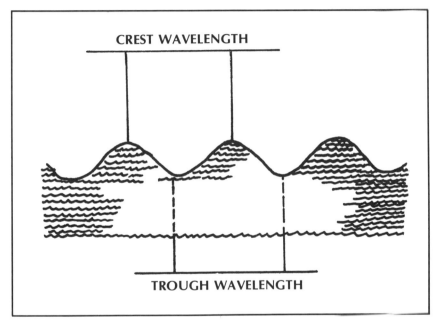

A wave is a disturbance traveling through a medium, or in the case of electromagnetic radiation, an accelerating charge that does not require a physical medium to move. The length of a wave is determined by measuring the distance between two crests (high points) or troughs (low points) as it passes a given position. Although a wave may travel a long distance—the trillions upon trillions of miles from a star in deep space, for instance—the wavelengths of light are very small. Violet light, for example, has a wavelength of only 1/65,000 of an inch (that means you could get 65,000 wavelengths per inch). (Credit: Betty L. Wright)

An electron in the excited state is like a heavy safe teetering on the ledge of a high building. It tends to return to the ground state level as soon as it can. When it drops back to its natural level at a lower orbit, it releases the energy used to push it up earlier. This energy is refunded or released as light. As we shall soon see, all of this is very important to the invention of the laser.

Because atoms differ in the number of shells available and the space between them, the amount of energy necessary to move an electron from a lower to a higher level in a given atom—of hydrogen, say—is specific and precise. For this same reason, the amount of energy refunded when an electron drops back to the ground state is also individually exact for a given atom. This means that every substance gives off and absorbs energy only in certain wavelengths. Thus, every atom releases a light spectrum that identifies it as clearly as a fingerprint.

Bohr's blueprint of the atom worked so well that he was awarded the 1922 Nobel Prize for his theory, but later scientists have refined his ideas steadily, providing an even more accurate picture of the atom.

These refinements included those of the French physicist Louis Victor de Broglie, who showed that not only radiation has the dual qualities of waves and particles, but that particles of matter also exhibit wave properties. His theory predicted that electrons of moderate speed would have wavelengths of about 1.65 Å. The similarity of the behavior of his predicted electrons to the behavior of x-rays at this wavelength verified his ideas. Later, the German physicist Erwin Schrodinger used de Broglie's theory of the wave behavior of particles to construct a mathematical description of wave mechanics. He suggested that the electrons swing all the way around an atom in a continuous wave rather than orbit it like a planet circling a star. This, Schrodinger explained, would account for the set orbits or shells in the Bohr model of the atom.

But the most important change in Bohr's theory (at least in terms of lasers) came from Albert Einstein in 1917. Einstein pointed out that Bohr's model of the atom explained only spontaneous emission of photons. While this accounts for up to 90 percent of all the photons given off by atoms, Bohr had missed one thing going on that Einstein's mathematical calculations revealed. In addition to spontaneous emission, Einstein said, a number of electrons can be "kicked" from higher to lower states rather than allowed to fall normally. This produced what is now called *stimulated emission* of light photons.

Stimulated emission is the atomic phenomenon that makes laser light possible.

Chapter 3

Masers to Lasers: How They Work and the Men Who Made Them

LASER MEN

When Einstein recognized the phenomenon of stimulated emission in 1917, he did not know that someday it would become the key to making a light brighter than the sun. Neither did anyone else, at the time.

Bohr's model of the atom explained how it absorbed and radiated energy most of the time. Tuned to a specific frequency spectrum, a given atom would take in and refund only energy of that spectrum and ignore others. If struck by energy of the proper frequency, red light, say, the atom would become excited, with electrons boosted to higher levels until they spontaneously dropped back to a ground state, returning the energy in exactly the same frequency and amount as that absorbed. But what happens when an already excited atom is struck by energy of the proper frequency?

This was the question Einstein asked. His answer: If radiation of the proper frequency strikes an atom already excited by that same radiation, the atom spits out its stored energy immediately. This is *stimulated emission*.

Stimulated emission amplifies (increases) the amount of light coming from the atom. An excited electron that plunges to its ground state spontaneously (which occurs in millionths of a second) would refund only the photon that originally excited it. If smacked by another photon of the same frequency while still excited, however, it returns both the photon which originally excited it and the one that kicked it prematurely to the ground state. Thus, two photons of the same frequency emerge.

A decade later, the brilliant English physicist Paul Dirac discovered that light produced by stimulated emission possessed an unusual quality. The two photons shot from the atom by stimulated emission emerged with their waves locked in step more precisely than the feet of a well-trained marching band. Unlike ordinary light with its undisciplined waves wiggling like thousands of disco dancers listening to different records, this light was *coherent*.

It was not until the 1950s that scientists realized that stimulated emission of coherent radiation could be put to use in remarkable ways.

CHARLES TOWNES AND THE MASER

On a pleasant spring morning in 1951, Charles H. Townes sat on a public park bench in Washington, D.C., making calculations on the back of an envelope. The numbers he wrote added up to one of the major discoveries of this century: a way to use atoms and molecules as amplifiers of radiation in frequencies above the radio and radar frequencies.

Often characterized as a Renaissance man, Townes has an impressive record of accomplishments in and out of science. He earned a degree in modern languages as well as a Ph.D. in physics, and speaks four foreign languages fluently. He has been a top executive with major corporations, a church deacon, and a Boy Scout leader. His activities include mountain climbing, skin diving, worldwide travel and teaching, as well as scientific research and invention.

During World War II, Townes worked at Bell Labs, where he became an expert in microwave communications and radar. Lying between the ultra-high-frequency radio waves and infrared regions on the electromagnetic spectrum, microwaves range from a frequency of one billion cycles (1,000 megacycles) per second to about 100 billion (100,000 megacycles) a second. This corresponds to wavelengths of about 30 centimeters (slightly less than one foot) to .3 centimeters (about a tenth of an inch). The dividing lines of the electromagnetic spectrum are not exact, however, and other frequency/wavelength figures extending a little way to each side are sometimes given.

The communications revolution of the late 19th and first half of the 20th centuries enabled radio scientists to produce coherent waves capable of carrying messages or information in most of the radio bands. By World War II, communications technology had climbed to the microwave region, though not deeply, to produce radar.

Charles Townes and his associate, James P. Gordon, with the first maser. Although this device looks like something from a 1950s science fiction television show, it paved the way to development of the laser. (Credit: Columbia University News Bureau)

Microwaves are particularly useful for this *Radio Detection and Ranging (radar)*, for at their high frequencies they travel great distances without spreading out like lower frequency radio waves. Radar engineers, including Townes, sought methods to go to higher and higher frequencies during the war to combat jamming. Sets used on aircraft sent out ranging beams at frequencies of 10,000 megacycles, at wavelengths of about three centimeters (a bit more than one inch). Townes worked on developing equipment that operated at the substantially higher 24,000 megacycle frequency, one-and-three-quarter centimeter wavelength microwave region.

While this proved unsatisfactory for radar uses, it interested Townes in microwave spectrography, which in turn eventually led to the invention of the maser and the laser. By 1951, Townes, now a professor of physics at Columbia University, had known for several years that resonances in atoms or molecules themselves might act as radio circuit elements. In the millimeter and submillimeter region and beyond, it becomes almost impossible to construct resonators to amplify the signal because the wavelengths are so short (a millimeter is about four-thousandths of an inch). At Columbia University's Radiation Laboratory, Townes made pioneering studies of the interactions between microwaves and molecules, particularly ammonia.

This work was supposed to find a practical way to extend radio technology into the very short end of the microwave region and above. Doing so had proved frustrating and continually beyond reach. On that warm spring day in 1951, Townes, discouraged by his lack of success, was in Washington to attend a committee meeting of scientists and engineers that the Office of Naval Research had formed to study the problem.

Townes, sharing a hotel room with his brother-in-law, Dr. Arthur Schawlow, was used to rising early because he had young children. Schawlow, then a bachelor, liked to sleep late, so when Townes awoke, he dressed quietly and went outside to enjoy the spring weather in the grey Washington dawn. "I found myself sitting on a bench in Franklin Park admiring the azaleas, then at the height of their bloom, but also wondering whether there was a real key to the production of very short electromagnetic waves," he remembers. *Light bulb!*

He suddenly thought of a way it might be possible to use molecules to generate the elusive very short radio waves. "In a few minutes," Townes recalled, "I had calculated—on the usual back of an envelope—the critical condition for oscillation in terms of the number of excited molecules which must be supplied and the maximum losses allowable in the cavity." His hastily scribbled calculations centered on the ammonia molecule. Dr. Schawlow has stated, "A detailed understanding of this molecule was just what Townes had needed to invent the maser."

Ammonia . . . that smelly stuff that is such a good solvent that it's used in cleaners of all sorts? Well, yes. Townes had studied the ammonia molecule because it has two specific energy states, the ground state and one upper level. The difference between them equaled the microwave energy frequency of 24,000 megacycles per second. (A molecule of ammonia, by the way, is a combination of different atoms: one nitrogen atom and three hydrogen atoms—chemists express this as NH_3).

What Townes wanted to do was to somehow excite most of the ammonia molecules in a container, then shoot them with a beam of microwaves at the frequency they absorbed so strongly, 24,000 megacycles. This would kick the excited molecules back to the ground state and produce a stream of coherent microwaves at that frequency.

The problem was how to push more molecules into the excited state than were in the ground state, producing what is called *population inversion*. Obtaining this topsy-turvy state proved difficult. With the help of his associates, James P. Gordon and Herbert J. Zeiger, Townes began working on a device to do it. The idea had occurred to him on April 26, 1951. He drew rough sketches labeled "apparatus for obtaining short microwaves from excited atomic or molecular systems" in his notebook on May 11. In early 1954, his research team obtained the first oscillations from an ammonia beam device. It was dubbed a *maser*, for Microwave Amplification by Stimulated Emission of Radiation.

Townes jokes that the acronym could also refer to: Means of Acquiring Support for Expensive Research.

HOW THE AMMONIA MASER WORKS

The maser Townes built consisted of several parts. The first, a cylinder containing electrically charged rods that attracted the low energy state ammonia molecules, but repelled those in an excited state. When ammonia gas spewed into the first chamber, the excited molecules repelled by the charged rods poured into an adjoining collection chamber. Thus, the device created a population inversion. This collection chamber, only centimeters long, was about the same size as the wavelength of the microwave photons. It served as a resonant cavity, in which the photons echoed back and forth from the walls, making sure all of the excited molecules were slammed back to the ground state, releasing additional photons in the process.

A device that produces a coherent wave is called an oscillator. The ammonia maser was the first oscillator that could beam coherent radiation at 23.87 gigahertz (23,870,000,000 cycles per second) streaming from a waveguide pipeline. By attaching a second waveguide to the collection chamber and beaming in a microwave signal at the proper frequency, the device could be used as an amplifier. The incoming microwaves joined with those produced by the dive-bombing ammonia molecules to create a much stronger, outgoing, coherent beam.

ATOMIC CLOCKS

Although the first maser demonstrated that atoms and molecules can be used to make coherent waves at higher frequencies, it had drawbacks. For communications, which requires tunability, the maser's frequency was too pure. A glance at your radio dial shows that the amplifiers it contains act on a broad range of frequencies (the numbers on the dial represent radio frequencies; the AM band in kilohertz, [thousands of cycles per second], the FM in megahertz [millions of cycles per second]). The ammonia maser, though, has such a narrow band width that it ignores incoming signals just a little off its precise frequency. But this frequency stability was also the ammonia maser's most useful quality, for it made the atomic clock possible.

Prior to the maser atomic clock, the world's best timepieces made errors amounting to about a second a decade. While that's more than accurate enough for ordinary purposes, even tiny errors can throw off the ultra-precise measurements that astronomers must make to track the stars or physicists and other scientists must use in their experiments. The ammonia maser, by oscillating at a steady, almost never varying frequency of nearly 24,000 million cycles a second, supplied a new standard. Atomic maser clocks accumulate errors at the rate of only one second in 10,000 years, a thousandfold improvement. Also, since each of the maser's 24 billion cycles each second acts as a tick, it gives scientists a scale of time measurement far beyond that of conventional clocks.

BLOEMBERGEN—SOLID STATE MASERS

The amount of power a maser can turn out depends on how close the molecules are in the substance in which population inversion is created. With a gas, such as ammonia, the molecules are few and distantly spread compared to a solid. Fewer molecules means fewer photons, and the power of the beam depends on the number of photons it contains.

So, researchers began looking for ways to make masers from solids, in which molecules are tightly packed. A number of scientists—Townes, the Russians N.G. Basov and A.M. Prokhorov, and Nicolaas Bloembergen at Harvard University—explored the possibility of using *paramagnetic* materials in masers. These substances are affected by magnetic fields, but not as much as truly magnetic metals such as iron. These are materials like sodium, potassium or the chromium in ruby.

These materials looked like attractive candidates because scientists believed, correctly, that they could take advantage of their atomic structure. In all atoms, electrons not only travel around the nucleus like the earth rotating the sun, they also spin like tops. Just as the night-day rotation of the earth around its own axis causes our planet's magnetic field, so too the spinning electrons create a magnetic effect.

However, most atoms have electrons married in pairs that spin in opposite directions, canceling the magnetic effect. Paramagnetic substances have atoms with some unbalanced electrons. This gives the substances their magnetic qualities. One electron of a pair is, in a sense, missing, so the spin of the electron which is present, and its magnetic field, are not counteracted.

According to quantum theory, atoms with these "free" electrons have energy states associated not only with the electron's level or orbit shell, but also with its spin. Since this free electron is itself much like a tiny magnet, it reacts to the presence of an external magnetic field. Scientists usually talk about what happens to the electron in a magnetic field in terms of angular momentum and other fairly complex mathematical language. (For those who would like to explore the nitty-gritty details we have included references in the bibliography.) Basically, however, what happens is that in certain substances, a magnetic field can split the energy level into two states—our old friends, the ground state and the excited state.

In some paramagnetic materials, the difference in energy states caused by the presence of an external magnet are in the microwave frequency. Therefore, by "pumping" with an electromagnetic field of the proper frequency, scientists can shove more atoms into the excited state than are in the ground state—creating the population inversion necessary for maser action.

Solid-state masers have several advantages. Due to the way electrons interact with outside magnetic fields, a solid-state maser can be tuned by adjusting the strength of the magnetic field. It can be operated over a large number of microwave frequencies and with a larger, less sensitive band width than gas masers. Moreover, since paramagnetic substances are solid, their closely packed atoms produce many more photons when stimulated to drop from excited to lower states. This means the resulting beam is much stronger.

Though scientists knew all of this in theory in the mid-'50s, engineering a working solid-state maser based on these principles required hurdling a number of obstacles. Normal temperatures required too much power, so researchers supercooled the paramagnetic crystals they used in a bath of liquid helium (−268.9 degrees centi-grade, which is only a few degrees above absolute zero). This super-cooling drains energy from the crystal's atoms, icily weighing them down to the ground state. With all of the atom's unbalanced electrons at the bottom energy level prior to "pumping," it becomes easier to make them behave as desired.

The first solid-state masers had serious limitations. Because the paramagnetic materials used had only two energy levels, an upper state and a ground state, the masers gave only short bursts of coherent energy. Once pumped by a high-powered blast of microwaves, the electrons emitted their photons and dropped back to the ground state. With no more excited atoms in the high energy state, no more photons could be released until the electrons were again pumped up.

In 1956, Bloembergen suggested using three-level paramagnetic substances to eliminate this problem. Some of these materials, such as chromium, have more than one unbalanced electron. Since sub-stances with one unbalanced electron will produce two energy levels in the presence of a magnetic field, those with two should have three useful energy states, Bloembergen said.

Two years after Bloembergen's proposal, scientists at Bell Labs built the first continuous beam solid-state maser. Other researchers constructed improved versions. These devices relied on the addi-tional energy levels provided by substances such as chromium (which actually has three unbalanced electrons). Here, population inversion was created between the upper- and mid-energy levels rather than just lower and upper levels. Continuous operation of the maser is possible because of this intermediate level.

Soon, refrigerated solid-state masers, with their ability to amplify extremely weak signals by using atoms as resonators, were put to work. Applications included radar, astronomy, and satellite com-munications.

When both Townes and Bloembergen shared the Liebmann Award for their work in 1959, Townes had the ruby from his radioas-tronomy maser mounted and gave it to his wife. Leaving the cere-mony, Bloembergen's wife asked why he didn't do something like that from his maser. The scientist replied, "But my maser was made of cyanide, dear." Cyanide, of course, is a strong, dangerous poison.

RUBY RED GLOW

The ruby crystal Townes presented to his wife foreshadowed the next development in the laser story. Ruby worked in masers because of its

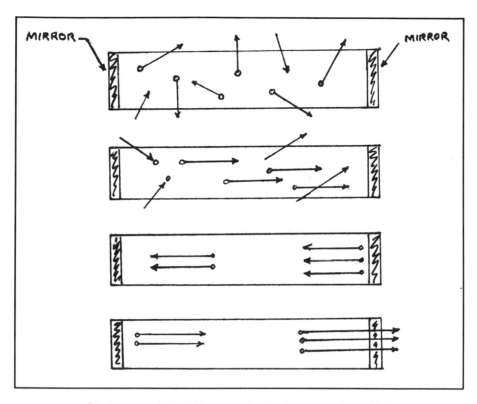

This is a simplified diagram of what happens after a flash-lamp excites the chromium atoms inside a ruby crystal. As some excited atoms spontaneously drop back to the ground state, they release photons of light. Those traveling parallel to the axis bounce back and forth between the ruby's mirrored ends, stimulating the emission of coherent photons by other excited atoms. Those not traveling parallel escape out the sides or are absorbed by the ruby. An army of photons traveling parallel multiplies at the speed of light, and eventually bursts from the partially mirrored end of the laser as a beam of coherent light. (Credit: Betty L. Wright)

chromium atoms. Actually a sapphire (which is a single crystal of aluminum oxide) with about five-hundredths of one percent chromium, the ruby glows naturally. Caused by the spontaneous emission of photons by the chromium *ions* (atoms with an electric charge caused by gaining or losing electrons or protons), this natural fluorescence makes it appealing as a gemstone. But its glow attracted scientists for a different reason.

"How was it that so many people at nearly the same time began to study the optical properties of the ruby?" asks Dr. Schawlow, who, with Townes, would later formulate ideas leading to the optical maser—the laser. One reason, he notes, is that advances in the understanding of these crystals suggested master principles might be "carried over" to optical frequencies. "I was intrigued by the fact that strong, sharp-line fluorescence in the deep red could be excited by broadband (bright, but incoherent) light in the green and blue region of the spectrum," said Schawlow.

One of the major reasons rubies were studied, in addition to their glamorous glow, was simple. Since the ruby was being used in microwave lasers (masers), many were available at research labs. "It was possible," notes Schawlow, "to visit Bell Labs and find a drawer full of rubies from which you could easily borrow samples." Schawlow borrowed so many for his work that a Bell scientist kiddingly marked one experiment, "from ruby ALS (Arthur L. Schawlow) bought!"

Schawlow had begun thinking about extending the maser principle to shorter wavelengths—those of visible light—as early as 1957. "I wanted to identify the difficulties and see if solutions could be found for them," he recalls. At first, he had only "vague notions about using ions in crystals." He mentioned this to Townes, then a Bell Labs consultant, and they decided to work on the problem together.

They knew that incoherent light from a lamp, if bright enough, could be used to pump three-level or four-level optical masers. The output wave's coherence would be decided by the photons emitted by the resonating atoms, not the light source. However, the problem, yet again, was in creating a population inversion, forcing enough atoms out of the ground state they consider home and into an excited state. There was another difficulty, as well. In the optical region, excited atoms quickly lose their stored energy through spontaneous emission.

Townes and Schawlow studied the maser equation, which owed its existence to that spring morning in 1951. This revealed an important fact that would help them design the basic laser resonator, the device which would make the laser possible. They discovered that a light wave in a resonator would lose energy only at the walls, and would gain it by stimulating photon emission as it passed from one end to the other. Later, Schawlow realized that there might be a simple way to make an optical resonator. "We could remove almost all of the walls of the box, and leave only two small mirror-like sections facing each other at the ends of a long, pencil-like column of amplifying material," [such as a ruby rod] Schawlow has explained. "As long as the end surfaces were much larger than a wavelength, they would act as good mirrors and reflect waves straight back and forth between them." A wave angled even slightly off the straight axis would skip out the walls, missing the mirror. The remaining waves, passing back and forth between the mirrors many times, would be amplified.

Townes and Schawlow published their ideas in the December 15, 1958 *Physical Review*. "It attracted considerable interest," says Schawlow. "And some very plausible arguments were advanced to prove that it would not work."

A NEW LIGHT

Despite the initial skepticism, once Townes and Schawlow explained how to make a resonator by using a mirrored crystal, scientists went to work in labs throughout the world trying to make a device that would give man control of light. And the resonator proved to be a major piece of the puzzle—but not the only one.

Dr. Schawlow, for instance, tried out his own idea, attempting to get laser action "from a rod of dark ruby using a small flashlamp." His attempt failed, as did many others. Yet, as Schawlow noted later, in September, 1959, he had described "rather concretely the structure of an optical maser." He continued, "Well, if we knew the material and the structure, why not do it? Now, when we know the construction of a laser can be so easy, it is hard to answer that question. But when we did not know how to assess the difficulties, and since it had never been done, we believed that they might be formidable."

Then, in May 1960, Hughes Aircraft scientist Dr. Theodore Maiman put an end to skepticism. He achieved lasing action in a ruby crystal rod, creating the long-sought new light.

Why did Maiman succeed when so many others had failed? The answer is simple: he tried harder. Maiman excited the chromium atoms in his synthetic ruby rod with a brilliant flash of white light that sent 2,000 joules of energy coursing through the crystal—enough energy to boost a safe to the roof of a house.

Scientists measure energy in several ways. Brightness is measured in watts. This represents *how much* energy is radiated. But brightness alone does not describe the work energy can do. A candle is not very bright. You can rapidly pass your hand through a candle flame without burning your skin. Hold your hand over the flame for some time, however, and it can cause severe burns. Sitting on the beach under a bright July sun can warm you, tan you or burn you— depending on how long you stay under its rays. The work energy can

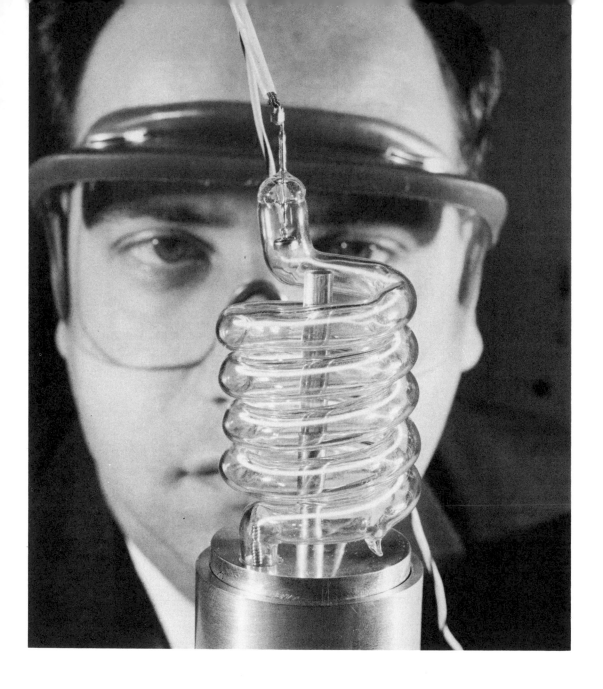

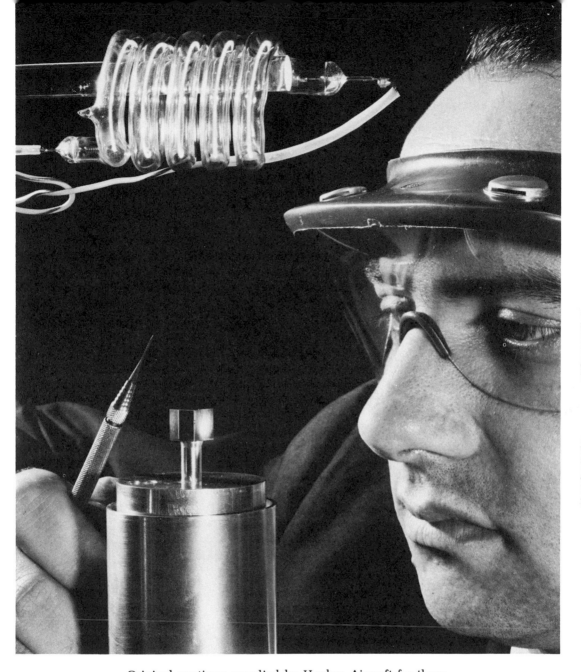

Original captions supplied by Hughes Aircraft for these historical photographs of Dr. Theodore Maiman with the world's first laser quoted Gray's Elegy, stating that "Full many a gem of purest ray serene" had electronic age implications never intended by the poet. Calling the laser's coherent beam an "Atomic Radio Light" of the purest colors ever known, Maiman accurately predicted the laser would "light" the way for improved communications and find a host of other uses.

do, then, is measured in power (brightness in the case of light) multiplied by time (*how long* it radiates). A watt is one joule second, or one watt of power radiated for one second. Climbing a flight of stairs requires about 1.66 joules of energy.

A thousandth of a second after Maiman tripped the 2,000 joule flashlamp (pouring enough energy to climb more than 1,200 flights of stairs into his ruby), pulses of red light streamed from the laser. Each bolt lasted a millionth of a second, and in two thousandths of a second, the entire process was over.

SOMETHING TRULY NEW

That tiny parcel of time was enough for something truly new to enter the world. At peak power, the spikes of laser light radiated 30 kilowatts (30,000 watts) in a one square centimeter beam, probably the brightest light ever seen in the solar system. The sun itself only radiates seven kilowatts per square centimeter at its surface. An ordinary household light bulb may range from 20 to 250 watts, with 100 perhaps the most common.

Yet, in terms of working power, Maiman's laser did not have much to offer. Despite the tremendous amount of energy he pumped in, the pulses of laser light he got in return were so brief that even at peak power they delivered only from one to three tenths of a joule, or about one twenty thousandth of the energy pumped in. But those bright fingers of coherent light pointed swiftly, and just as certainly, to the future.

Even in press releases decades old, you can feel the excitement and sense the atmosphere that something very important was happening. Let's look back at July 7, 1960, New York City's Hotel Delmonico, and recapture a taste of Maiman's heady pleasure in announcing his discovery.

It is morning. Maiman, then the youthful director of the Hughes quantum electronics department, stands before the assembled science writers of the press. His Ph.D. in physics from Stanford University is only five years old, but he is about to announce that the "long-sought laser is no longer an elusive dream, but an established fact."

Showing the audience his invention, a thin ruby rod, surrounded by the spiral coils of a flashlamp, the whole thing about the size of a glass tumbler, he passes it among them as he explains its significance.

"One way of explaining the importance of the laser," Maiman offers, "is to say that as a scientific advance it projects the radio spectrum into a range some ten thousand times higher than that previously attainable. It marks the opening of an entirely new era in electronics.

"The radio spectrum . . . is the range of electromagnetic frequencies starting with commercial radio at one million cycles per second and extending into the upper microwave region of fifty thousand million cycles per second. The laser jumps the gap to five hundred thousand *billion* cycles per second, opening the way for many important applications. For the first time in scientific history, we have achieved true amplification of light waves."

In the 15 years following World War II, Maiman notes, man extended his ability to use frequencies of the electromagnetic spectrum by only a factor of five. The laser, he asserts, "represents a jump by a factor of ten thousand." Because the laser uses atomic methods to generate light and its coherent properties are similar to those of radio waves, he describes it as "atomic radio light." Apt though this name may be, it will not catch on; the laser, however, will, and Maiman himself explained why.

"As a direct consequence of its coherence, the laser is a source of very high 'effective' temperatures . . . higher than at the center of the sun or stars."

An effective or "equivalent" temperature means the heat at which an ordinary light source would radiate a signal as bright as the laser's at the laser's color, he explains. To vie with a laser, a Hollywood klieg light would have to reach a temperature of several billion degrees—an impossibility, since it would burn up long before reaching that level. The laser beam achieves these effective temperatures by concentrating its energy to a pinpoint. This raises the temperature of the target rather than of the device producing the beam (although some lasers do require supercoding because of the tremendous power being pumped in).

What can such a beam accomplish? On that day in 1960, Maiman suggested many applications that have since become a reality. These include radar and space communications, investigation of matter and other scientific research, providing communications channels in unprecedented numbers, and medical, industrial and chemical purposes.

FROM LASER . . . TO LASER . . . TO LASER . . .

Within months of Maiman's announcement, other research labs had made working lasers. By the end of 1960, at least four other lasers were operating. At IBM, Peter Sorokin obtained laser action from trivalent uranium ions and other "four-level" substances. Just as masers that used atoms with extra-energy-state levels produced stronger signals, lasers based on atoms with four rather than two energy levels got "extra help." Exciting electrons to the fourth level made it easier to keep a steady population inversion at lower levels.

Bell Laboratory scientists Ali Javan and others achieved laser action from a gas, helium-neon (see Listen to the Light—Lasers in Communications). This was the first continuous-wave laser as well as the first laser to use a gas. It operated at invisible infrared wavelengths, as did the IBM laser.

Almost too fast to keep up with, the field continued to expand. New substances were made to lase, while old ones, such as ruby, were refined through improved methods of growing crystals. New methods of pumping, focusing and increasing the power of lasers were introduced. Before 1960 ended, five different kinds of lasers existed.

Another Hughes researcher, R.W. Hellworth, found a trick to increase laser power dramatically only months after Maiman announced the first ruby device. Hellworth discovered that a fast shutter like that in camera lenses could trap atoms at their excited state in a laser resonator. The shutter, blocking one of the resonator (ruby rod) mirrors as light pours in from a flashlamp, prevents atoms from zipping immediately back to the ground state. This way, stimulated emission of photons is prevented until the resonator cavity is bursting with atoms ready to dive-bomb to their ground state and refund the pumped-in energy as a coherent beam. At the last instant before the pumping flashlamp dies, the shutter snaps open. So many atoms are now ready to drop that stimulated emission and the resulting laser action occurs very quickly, and much more powerfully than otherwise.

Q-SPOILING

When the additionally amplified laser beam blasts from the rod, it is powerful enough to boil water a thousand miles away. Hellworth called this trick "Q-spoiling"; the "Q" stands for the quality of the resonant chamber, which the shutter momentarily spoils. Perhaps "Q-improving" would have been a more appropriate name. Later, rotating mirrors were used to perform the same power-boosting trick. This works because the mirrors must be exactly parallel to each other for stimulated emission to start. The process is sometimes referred to now as "Q-switching."

Kicking a laser with a laser proved another effective method of amplifying the power of a beam. Using a ruby rod without mirrored ends, researchers pumped its chromium atoms to their excited state with a bank of flashlamps. Then, they fired a Q-switched ruby device at the first rod. The superpowered spear of light kicked into the excited atoms, picking up even more intensity before emerging.

These and other means made ruby lasers with peak spikes of 1,500 joules and 500 million watts possible in pulsed bursts of billionths of a second. Major cities do not require this much power during a given instant. Lasers using 6- and 12-inch rubies reached 350 to 500 joules. Many of these units required supercool baths of liquid helium or nitrogen (at minus 273° C) to carry off the immense heat they generated.

Ten years later, scientist Peter Franken recalled what the early days of laser development were like. Speaking to the Optical Society of America in 1971, he said: "I recall now . . . the first meeting of any professional society on the laser, the spring meeting in 1961 of this Society. At most meetings, people carry some cameras, and a man will show a slide, and you'll hear a few clicks. At that meeting, every time a slide was projected, the cameras sounded like machine guns. My good friend Arthur Schawlow was talking about some of the puzzles and mechanisms of the ruby laser.

"He went to the blackboard, picked up a piece of chalk, and wrote down the number one, turned away from the blackboard and a dozen cameras went off."

Dr. Schawlow himself has noted that "From the beginning, writers of popular accounts in newspapers and some people in the military expected, or at least hoped, that lasers would fulfill the old dream of a 'death-ray.' " This factor created some of the excitement about lasers, though then, as now, Schawlow was skeptical about their use as weapons.

"We had a good bit of fun with this notion," he wrote in *From Maser to Laser*. "I have often shown a slide of our death-ray countermeasures, which I made at that time. It shows some suits of shining armor, of the kind that knights used to wear. It was easy to calculate that a two-hundred-pound man could be completely evaporated by about 200 million one-joule shots from a typical ruby laser. If we could deliver them at the rate of one per second, which was rather better than we could do, he (the knight) would only have to stand there for six years."

Still, Schawlow noted that "once the principle was established," large, high-power, continuous-wave lasers were virtually inevitable. Indeed, in a later paper for a scientific journal, after many advances in

Better to light a candle than curse the darkness. Dr. Arthur Schawlow, who, with Dr. Townes, published the first scientific paper describing how a laser might be made, is shown here lighting a candle with a low power laser. (Credit: Stanford University)

laser development, Schawlow wrote: "Quite marvelous and unexpected things have been discovered. Yet I cannot help thinking that even greater surprises may lie in the future."

Gordon Gould and the Laser Patent Maze

Science often involves drama, controversy and human lives as well as flashing lights and columns of numbers. The laser story includes a prime example in Gordon Gould's controversial fight for recognition as inventor of many basic laser components.

After a 20-year trip through a twisting maze of patent battles, Gould realized a long held ambition: on October 11, 1977, he was issued U.S. Patent No. 4,053,845, entitled "Optically Pumped Laser Amplifiers." If upheld in court, the patent could give him royalty rights affecting as much as a third of the laser industry.

In April, 1979, Gould was awarded a second laser patent that could give him rights to use of a laser system to spark fusion energy. Two more patents covering other aspects of lasers are pending. All, as of 1981, were still being fought in court by industry giants such as Bell Labs, Hughes Aircraft and Westinghouse.

But who is this Gordon Gould? How is it that he is being called "the forgotten inventor of the laser" in some magazine articles?

Gordon Gould claims that he not only invented several basic laser concepts, but also that he coined the word laser as well. A genial, easy-going man, Gould bases his claims on notebook entries he had notarized in a candy store in November, 1957. He tells it this way:

In 1957, Gould was a graduate student at Columbia University and had been working seven years on his Ph.D. thesis on interactions between light and atoms. "The atmosphere at the Columbia Radiation Lab was very fertile," Gould says. "Townes, who built the first maser and made a lot of other contributions, was there, and others. It was thriving with new developments, and the idea of the laser was very much in the air."

Gould had been working on the idea of optical pumping systems for years, even preceding his arrival at Columbia. He was particularly interested in the possibility of using light to excite thallium. Unable to fall asleep one Saturday night in early Novermber 1957, he thought about the laser problem. "The idea came in a flash," he says. "I spent the previous 25 years of my life getting ready for that moment. I thought of a configuration for generating a laser beam and spent the rest of the weekend writing it down."

The following Friday, he took the gray notebook containing his ideas to a Bronx candy store owner who was also a notary. The first page of the notarized entries reads: "Some rough calculations on the

After decades of struggle, inventor Gordon Gould won patent victories that may make him a rich man if they are upheld in court. (Credit: Glen Gould)

feasibility of a LASER: Light Amplification by Stimulated Emission of Radiation." It included a crude sketch and description of a possible laser resonator with reflecting, parallel mirrors.

Several weeks later, Dr. Townes called Gould and asked about the thallium lamp he was working on. The nature of Townes's questions suggested he might be considering the same ideas for a laser, Gould recalls. Gould went to a patent attorney. The lawyer told him, incorrectly, that he would have to build a working model to get a patent. Gould dropped his Ph.D. work, to which he never returned, and joined a company known as TRG to attempt to build a working laser.

There he conceived other laser ideas, which he recorded in another notebook in December 1958. These included suggestions that a laser could produce the intensive heating which might make it a formidable weapon, industrial tool or fusion power igniter. In 1959, the Defense Department awarded TRG a one-million-dollar contract based on these proposals. Nevertheless, at this point Gould had neither published his ideas in a scientific journal nor built a working laser.

In addition, his connection with the Defense Department and a bit of his past led to a serious difficulty. At the beginning of World War II, Gould, with the ink still moist on his bachelor's degree, joined the Manhattan Project, the U.S. war effort to construct an atom bomb. At about the same time, he and his wife joined a Marxist study group. This act cost him his security clearance and he was dismissed from the Manhattan Project despite the fact that he soon quit the study group.

Sixteen years after his 1943 attendance at the Marxist meetings, the incident came back to haunt him again. The Defense Department barred him from working on his own laser project at TRG, and even denied him access to his own notebooks, which were classified as government secrets. (One of the many ironies involved in Gould's story is that his membership in the Marxist group was discovered because the leader was an FBI agent.)

The road to scientific credit for an idea is to publish in a respectable journal. In December 1958, Townes and Schawlow published the first description of a laser, and in July 1959, they filed for the first patent on the device. Gould's claim, filed nine months later, met with 17 years of setbacks in patent ''interferences'' and appeals. But American patent law, as Dr. Schawlow notes, is ''strange.''

In most countries, the first applicant to file a patent is legally considered the first inventor. In the U.S., a later applicant can claim that title—legally, anyway—if he shows prior conception. This requires such things as documented proof, a notarized notebook, reduction to practice (making it work) and, in some cases, showing diligence (hanging in there and trying to make it work). Gould lost all the early rounds of lengthy battles in patent courts to claims by Schawlow, Townes, Javan and Hellworth.

Appeals courts ruled against Gould's interference claims time and again, ruling that his notebook entries lacked critical information in one case, and that he had not shown the ''diligence'' required by law. In the Javan gas-laser case, the courts decided Gould's description of such a device would not permit an ordinary worker in the art to build one—a decision influenced by the fact that TRG, to whom Gould had assigned his patents, did not do so. Gould had also, in his notes, proposed a Q-switched laser, but the court ruled against his claim here too, again apparently influenced by TRG's inability to construct one.

Discouraged and in debt, but still trying, in 1975 Gould gained a new partner in his fight, the Refac Technology Development Corp. Refac, with the aid of a team of professional patent attorneys, reversed the trend in 1977. Although other basic patents in the laser field recently expired, if Gould's holds up, the rights will be worth a fortune. He has already received considerable sums for shares in those rights.

"I'm not bitter. How could I be after so many years?" Gould said when his first patent was issued. Now vice president of a fiber optics and laser engineering firm, Optelcom, he says he looks on it all now as "just an interesting adventure." And besides, he may be in line for as much as seven million dollars annually in royalties.

DID HE OR DIDN'T HE

"A patent is only a license to sue," says Dr. Schawlow, who remains skeptical of some of Gould's claims and bemused by U.S. patent law. "I would like to be more gracious to Mr. Gould. I think he probably did realize that you could get high power out of lasers before we did. We really didn't think about that. But I'm not very objective."

While noting that it's now a matter for the courts to decide, Schawlow disputes some of Gould's claims, as does Townes. Schawlow also draws a distinction between the way things are usually done in patent law in the U.S. and the way they are done in the scientific community. "The scientific community usually recognizes the first person to publish. That takes a little courage, to put your name on something, to say I believe this. If it's wrong, people will think you're a fool. In patent law, the first person who records it in a notebook or anything properly witnessed can claim priority for the patent. We did what we did and we published it. Mr. Gould kept something in his notebook, and years later it emerges. I think the issue is whether he found something that we failed to state explicitly enough.

"He might be able to get rich on something we thought was too obvious to mention. It sounds like what happened when Watt invented the steam engine. Somebody else patented the crank he used. That's why he had to invent a crazy gear motion.

"The courts will decide whether he anticipated us in any way, but remember that lasers went ahead and got built without any help from Mr. Gould at all."

Chapter 4

The Death Ray:
Lasers as Weapons

WW III?

A future scenario: the next war?

The war began in space. Initial skirmishes for control of this ultimate high ground, fought with briefly flashing rapiers of laser light, invisible rays and silent explosions, have already occurred. Hunter/Killer satellites of both sides found and destroyed their targets—other orbiting spacecraft used for military communications. In engagements taking only fractions of a second, armed satellites crossed laser swords, turning each other into so much space Swiss cheese.

As Soviet missiles blast from silos in the frozen Siberian tundra and race toward Western cities, another space battle is still in progress. United States Orbiting Laser Defense Systems (OLDS) struggle for survival. Their defensive lasers detonate Soviet Hunter/Killers before they are close enough to do damage. Shifting mirrors deflect beams aimed at the stations from mountain-top-based Soviet laser cannons.

Tracking the Soviet missiles, the OLDS stations prepare to fire while the boosters are still in their most vulnerable stage, the first 50 seconds after launch. Seed lasers lock on target. Although the Intercontinental Ballistic Missiles (ICBMs) move at seven times the speed of sound (more than 5,300 miles an hour) this is no match for laser beams zipping through space at the speed of light (186,000 miles a *second*). A missile moves only inches from the time the seed laser begins tracking to the time a powerful weapons laser strikes.

Two types of orbital laser battle stations fight the Soviet attack. The first system, a fleet of three dozen hydrogen fluoride chemical lasers mounted on satellite platforms and focused by large, highly polished mirrors, fire repeatedly. Methodically tracking their targets

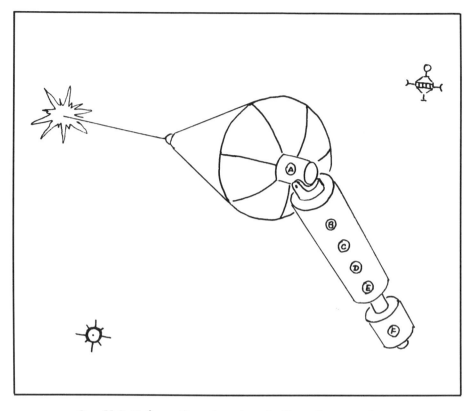

One U.S. Defense Department projection of a space laser weapon looks like this drawing. Its parts, from top to bottom (represented by lettered circles), include: (A) A sophisticated tracking/aiming system that must lock onto a target traveling 9,000 miles per hour. (B) The laser lenses that tighten an already coherent, narrow beam to deliver a powerful punch over long distances. (C) The laser amplifier, or fuel—tons of CO_2 gas or a substance that produces ultra-violet or X-ray light. (D) Microminiature sensors that detect signs that the missile target has been hit so that the laser can shift focus to the next most dangerous threat. (E) A communications system for control and information processing. (F) The satellite's engines and fuel, used to keep it in position or dodge killer satellite attacks. (Credit: Betty L. Wright)

with sophisticated equipment, they spit superheated energy beams at the nuclear missiles, holding on the booster's skin until it ruptures.

Again and again they lash out, switching to a target posing the next greatest threat as each previous ICBMs fuel and/or warhead explodes harmlessly high above the earth.

Some missiles, however, have been modified. They are not vulnerable to ordinary laser attack. They spin, preventing a laser from holding on a single spot, and have either highly reflective surfaces or protective coatings that peel away, dissipating laser energy harmlessly. As these penetrate the first laser defense system, the second tier comes into play. They are rare gas halide lasers arranged in rings of 50 around a nuclear device. Each laser locates a target and locks on through a seed laser. An instant later the nuclear bomb explodes, funneling a dense plasma through the laser beam that strikes the target with great force. There is no defense against this "impulse" kill. Few enemy missiles survive this gauntlet of light. Lasers, often touted as a death ray, instead save millions of lives in this instance.

Space war, swords of flashing light, invisible rays and soundless explosions—while all of this may seem like *Star Wars* science fiction, it isn't. The Pentagon and Soviet defense experts play out scenarios like the one you have just read to determine what kind of space-based laser defense will be most effective. It is no longer a question of *if* they will be used; it has become a matter of when and how they will be used.

The U.S. Defense Department has spent two billion dollars on laser weapons research in recent years, and announced increasing expenditures for the 1980s due to fears that the Soviet Union is far ahead in laser weapons development. The U.S. Army, Navy and Air Force all have separate laser weapons research programs. What amounts to laser ray guns are being developed not only for use in space, but also as ground and warship-based anti-aircraft and anti-missile defense, air-to-air and air-to-ground weapons.

Despite its multitude of constructive peacetime uses, the laser's potential as a weapon caught the imagination of the news media and the military from the day of its inception. Here, they thought, was another example of science catching up with science fiction. The laser appeared to be the "death ray," the Buck Rogers blaster, the real Flash Gordon stuff, with emphasis on the *flash*.

Initial research, however, quickly demonstrated that laser weapons presented a host of knotty engineering problems. For a time, some scientists suggested that all discussion of laser weapons was still sci-fi and likely to remain so in the foreseeable future. Many lasers, particularly early ruby crystal models, operated inefficiently. Only a tiny portion of the enormous energy pumped in to make them work was returned in the beam—less than one percent in many cases. The other 99 percent appeared as heat.

Without cooling, this heat would melt the nozzle of a laser powerful enough for weapons use. Although cryostats, special Thermos bottles containing supercooled liquid helium or nitrogen, can solve this problem in the laboratory, they are not handy battlefield equipment. Combined with the bulk of an energy source large enough to pump high-energy lasers, they make the instruments bulky, vulnerable and limited in mobility. They were a long way from the snappy, hand-held ray gun of Buck Rogers.

Also, using lasers in the atmosphere leads to a number of problems over long distances. Fog, mist, rain, clouds and smoke (among other things) spread or stop light . . . even laser light. Thermal blooming (heat waves around the laser beam) and atmospheric turbulence (air movement caused by temperature changes) can diffuse the beam, stealing its power.

With large grants of federal money, engineers, scientists and technicians in government labs and private industry went to work on these problems. Methods and materials to increase the power and efficiency of lasers while reducing their size were rapidly developed.

MORE POWER

Lasers using carbon dioxide proved interesting for weapons use because of their efficiency. It works like this: An electric charge is sent through a tube of nitrogen separate from carbon dioxide, raising the molecules of the gas to an excited state. Both the nitrogen and carbon dioxide then flow through an interaction chamber, where laser action begins. The result is a powerful laser beam emitted in the invisible infrared frequency.

One of the first lasers of this type, developed by the Canadian Defense Research Establishment in 1970, produced beams with peak powers of 100 million watts. It instantly vaporized substances such as steel, wood and asbestos. Another carbon dioxide laser, developed by the U.S. Air Force, looks like a rocket engine. It heats the two gases in one chamber to 3,000 degrees Fahrenheit, then blasts them through a nozzle at supersonic speeds into a second chamber. There, a portion of the carbon dioxide molecules lase, producing a beam powerful enough to zap air-to-air missiles in flight. These lasers have efficiencies of 5 to 10 percent, 40 times greater than other gas lasers and 9 times more than the ruby laser.

As the catalog of materials that could be made to lase grew, so, too, did possible military uses of the ever more powerful beams produced. Excimer rare gas halogen lasers (*see* further, pages 70–72) pumped by extremely high-energy electron beams are one of the most recent developments. They are attractive to the military because they operate at short ultraviolet wavelengths with 10 percent efficiency. Since the shorter the wavelength of a laser, the more energy it concentrates in its beam, these new generation "Uvasers" do more damage when they strike a target. In addition, smaller optics such as mirrors and lenses can be used to focus and aim them.

Lasers on the Battlefield

It is night. Icy rain whips against the faces of GIs manning a forward observer post. Kneeling in a puddle of muddy water, the soldiers set up their infrared telescope mounted over a laser.

In the distance, they hear the rumble of an enemy tank. Sighting through the night-penetrating scope, one of the observers centers the tank's image. He squeezes the trigger on the device—but not to fire a weapon. Instead, an invisible beam of light speeds to the tank and back at 186,000 miles a second.

The laser range finder's receiver registers the tank's exact range, azimuth and elevation. It automatically transmits the data to an artillery battery miles away. Flipping another switch on the device, the soldier marks the tank with a "designator beam" of coded laser light. Homing in on this guiding beam, a laser tracking missile or artillery shell wings to the target and blasts it. The enemy, even if they had sophisticated equipment to tell them they had been sighted, could not avoid their fate. The entire process takes only seconds.

Optics technology is another major area of laser weapons research. Just as one can focus the sun's rays through a magnifying glass to burn paper or char wood, lenses and concave mirrors can be used to tighten the already coherent, narrow beam of a laser to an even finer point. Lasers have been focused to a point tiny enough to drill 200 distinct holes in the head of a pin. Focusing the beam "bunches" the photons, concentrating the energy in even tighter packets, creating a hot needle capable of raising temperatures to 18,000° C at the spot on which it is focused. That's three times the temperature of the surface of the sun, and by no means marks the limit of what lasers can do.

Still another recent development, the free electron laser, is described as "one of the most exciting new concepts," by a U.S. Defense Department official. It combines particle beam accelerator and high energy laser technology to produce a "tunable" laser that operates in visible and ultraviolet wavelengths. They have an incredible efficiency of between 30 and 40 percent. Both the free electron and excimer lasers offer much greater range than previously obtainable, a factor of major importance in weapons applications.

New optical devices that vastly extend the range of laser weapons are also in development. The U.S. Defense Advanced Research Projects Agency (DARPA) concocted a device called a beam expander that works like an inverted telescope. Going in one end, the laser beam is expanded to the large diameter necessary for it to project great distances. Coming out the other end, it is aligned so that it will decrease as it travels and strike the target as a narrowly focused lethal ray. Combined with the new high power lasers, this technology extends effective beam range to as much as 8,000 kilometers (4,971.2 miles).

While laser weapons technology is by no means perfected, refinements of design and the recently discovered likelihood that Soviet ICBM boosters are more vulnerable to laser attack than previously thought, make lasers increasingly attractive to defense analysts as space-based weapons. Space is a hospitable battlefield for lasers, since they work more effectively in its supercold vacuum. And, the new-found vulnerability of Russian missile boosters means that lasers do not have to vaporize an ICBM to destroy it.

HIGH RISK

Soviet advances in combat laser technology are considerable, though. Some analysts assert that the Russians are years ahead of the U.S. in laser weapons development. Many military and government defense experts think it is absolutely necessary for the U.S. to play catch-up. Experts say now that with the aid of the Space Shuttle, anti-missile laser battle stations could be guarding the skies by the mid-1980s or early 1990s.

One U.S. Senator, Malcolm Wallop (R.-Wyo.), believes that "several dozen satellites equipped with high-energy lasers could provide an effective defense against an ICBM attack." A member of the Select Committee on Intelligence, the Senator told the Institute for Foreign Policy Analysis that experiments conducted with "containers similar to the boosters of ballistic missiles" demonstrated that they buckled or burst within a second of being hit by a laser. A 2-megawatt chemical laser using a 33-foot mirror or a 5-megawatt device with a 13-foot mirror would produce a beam intense enough to destroy the boosters

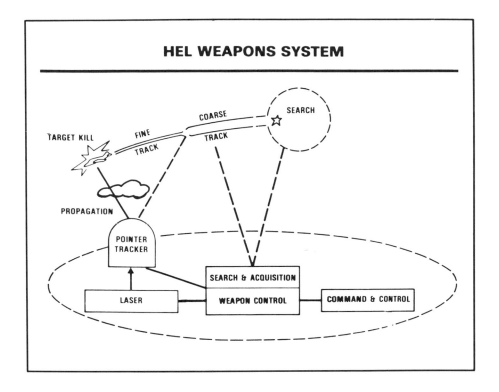

This Department of Defense diagram shows that there is more to using lasers as weapons than just pushing a button. Beginning at the left, a laser weapons system must search to find its target, track the target, overcome atmospheric problems (clouds, smoke, rain), and then stay on the target long enough to produce a "kill."

at a range of 4,000 miles. "Any nation which deployed three dozen of these first-generation chemical laser platforms in a low-altitude orbit would be able to deny the rockets of any other nation the privilege of entry to space," the Senator said.

Some military analysts, however, still argue that lasers are a high risk technology because of possible defensive measures the Soviets or other nations could take to protect ICBMs. Others contend that every defensive measure taken adds to the cost and detracts from the effectiveness of missiles and satellites.

Scene Two

Technicians in surgical garb hover over microscopes in a sealed room as clean as a hospital operating room. They are inspecting the components of a laser device; each device contains nearly 1,000 parts. Each part must be carefully cleaned and calibrated during assembly. Some are tiny and must be handled with special tweezers.

The parts are packed in vacuum-sealed plastic bags. Special ultraviolet and fiber-optic backlighting scans each part for almost invisible dust. Air movement in the room is kept constant and clean.

"These two scenes bracket a paradox," asserts the Hughes Aircraft publication, *Vectors*. The laser range finder Hughes constructs must survive rugged and messy battlefield conditions. But it is built under immaculate, extraordinarily controlled conditions in a sealed laboratory. The reason for this contrast is that no matter how hostile the battlefield environment, from swamp to desert, icefield to tropical jungle, internal contamination of laser devices are their biggest problem. These can be combated by constructing the devices under rigidly controlled conditions so that the internal parts will work regardless of the external environment.

The reason for this is that any kind of contaminant inside the laser device can be incinerated by the beam, pitting a lens or other optical part and ruining the equipment. Properly sealed and protected, however, the device will do a job no other range finder can, rain or shine.

To counter this argument that lasers are perhaps too easily defended against, DARPA is developing a number of alternative space based anti-missile programs. In at least one, no known defense would be available to the enemy. It suggests pumping a ring of orbiting lasers with a small nuclear bomb.

In this system, a fleet of laser platforms, each with 50 one-shot lasers arranged in a ring surrounding a nuclear device, would be orbited. Each laser would be individually aimed at a separate target. When the nuclear device explodes, it fuels the rare gas halide lasers, which in turn spew out a dense plasma beam that would knock out an ICBM by impulse—force of impact—rather than heating effects. The lasers would be limited to one shot because debris from the nuclear explosion would destroy them a thousandth of a second after they fire.

Soviet advances in laser technology during the last few years is spurring increased U.S. research into numerous space-based laser weapons, not just anti-missile platforms. The U.S.S.R. has tested killer satellites 17 times that U.S. analysts know about. One type sidles up to its victim and explodes, destroying the enemy satellite with a barrage of shrapnel. Others in development would use lasers. To combat these killer satellites, the U.S. is developing both anti-satellite and defensive satellite laser weapons.

Another reason space-based lasers are in the forefront of U.S. weapons research is stated by Senator Wallop. "They cannot harm anything on earth and are fit only to destroy weapons of mass destruction that travel in or on the edges of space," he said on the Senate floor. "These are the rarest of weapons in that they do not kill but only save lives."

DOWN TO EARTH

While space-based laser weapons are currently the major focus of U.S. research, other military applications of high energy beams are also being studied. The Navy, Air Force and Army all have programs underway to demonstrate the lethality of lasers against realistic air, ground and sea targets. The Navy has invested $200 million in high energy laser research since 1975, and the Air Force twice that amount. The Army spends about $1 million a year in its research.

The Air Force has a three-part laser technology development program conducted at the AF Weapons laboratory at Kirkland AFB, New Mexico. They include:

1. The Airborne laser laboratory, a highly instrumented Boeing NKC-135 aircraft used to demonstrate the usefulness of various lasers in combat situations. Present research centers on using a carbon dioxide dynamic laser to shoot down air-to-air and surface-to-air missiles.

2. Advanced technology designed to support laser weapons in areas such as optics, fuels and new ideas.

3. Development of advanced lasers such as chemical instruments and an oxygen-iodine weapon that operates at more effective shorter wavelengths (1.3 microns—near infrared).

"A laser delivers its photon bullets at the speed of light," one Air Force official told *Aviation Week and Space Technology* magazine. "That eliminates the need to lead the target," he explained. In today's world of supersonic missiles, interceptor aircraft and electronic warfare, military planes, ships and ground forces need every advantage they can get to survive.

A Hughes Aircraft technician uses a fiber optic cable like a tiny flashlight to examine the optics of the Army's Ground Laser Locator. Note the hospital-like clothing. Technicians wear rubber gloves and other special garments and work in a special "clean" room with a filtered air system to keep even the tiniest amounts of dust away from the sensitive laser components. (Credit: Hughes Aircraft)

The Navy Sea Lite program coupled a chemical laser device with a pointed/tracker system to blast Hughes TOW wire-guided antitank missiles and drone helicopters from the air. The TOW missiles pull several hundred gs (several hundred times the force of gravity) coming out of the launch tube, which makes them a very difficult target. Yet the Navy laser went four-for-four: four hits, no errors. As targets,

---

The Death Ray

"It is a matter of wonder how the Martians are able to slay men so swiftly and so silently . . . A beam of heat is the essence of the matter. Heat, and invisible, instead of visible light."

H.G. Wells, *War of the Worlds,* 1898

A poster with that quotation decorates the wall of physicist J. Richard Airey's Pentagon office. Head of the U.S. Directed Energy Technology Office, Airey is charged with trying to make science fiction a reality.

Airey designed one laser weapons system that the Navy tested. It blasted antitank missiles traveling at 450 miles per hour from the air. "You can't dodge a laser beam," he notes. "As soon as someone presses a button, the light hits the object."

The English born scientist finds himself in the middle regarding the "can we/can't we" laser weapons controversy. "No matter how hard we push," he believes, "it will be 1990 before we can build an effective space-based laser system with the complex control systems it needs." Yet this attitude, considered conservative in some circles, makes him a "wild-eyed proponent to those who think it can't be done at all," Airey says.

A friend once asked him, "How do you sleep at night?" after learning he worked on "such a terrible thing." The physicist replied: "It would be nice if we didn't have to have any weapons, but that's not going to happen. We need a strong defense, and somebody has to worry about it."

---

the Hughes missiles were much more demanding than the type of low-flying cruise missile or interceptor aircraft a laser system would have to fight at sea, Defense Department officials claim. Now the Navy is concentrating on scaling up the power of its lasers. It plans to test a 2.2 megawatt deuterium-fluoride continuous-wave laser operating at 3.8 microns (mid-infrared)—five times more powerful than any other model tested in the U.S.

The mini-laser range finder being installed on this Cobra attack helicopter will enable the chopper's gunners to direct rocket and cannon fire with deadly accuracy. (Credit: Hughes Aircraft)

Constructed by TRW company, the laser will be pitted against numerous realistic targets to determine whether scaled up power produces a practical laser weapon or "remains just flashy talk," according to a Pentagon official. It will be mated with a new optical system called "the beam director," the old pointing and tracking system upgraded by 18 optical elements between the beam and the director.

The Army's high-energy laser programs, while funded at lower levels than those of the Air Force, Navy and space-based programs, are nevertheless impressive. These programs include efforts to develop powerful air defense lasers for ground forces; lasers to blind enemy sensors on aircraft, air defense missile systems and tank laser range finders; and lasers used for large- and small-arms aiming and range finding. In the 1960s, the Army funded research at New York University in which lasers were used to kill small animals. The slaughtered animals were dissected to determine exactly what physiological damage occurred. In 1976, the Army used a carbon dioxide electric laser in a mobile test unit to slash both winged and helicopter drones from the sky. Currently, Army research is directed at developing practical high-energy lasers suitable for battlefield use. Lasers are frequently substituted for conventional weapons in Army war games to assess their effectiveness in terms of cost, performance and outcome of a battle.

While an honest-to-Buck Rogers laser sidearm has proved elusive so far, a so-called "laser squirt gun" has been proposed. It would use a chemical compound to fuel a ruby laser. To use the weapon, a soldier would hand-pump the chemical ingredients into the laser nozzle. There, they would flare, optically pumping the laser rod. Focused through a lens to killing intensity, the beam flashes for a fraction of a second—long enough to drill a hole through an enemy.

And, at least one hand weapon, a laser-sighted rapid-fire rifle developed by the American Research and Development Company, has earned the nickname "the Buck Rogers gun" among criminals and law enforcement personnel worldwide. A battery operated helium-neon laser mounted under the barrel as a sight gives the .22 caliber weapon unprecedented accuracy. Although it can fire up to 2,150 rounds a minute—enough to empty a full 177-round clip in five seconds—it is virtually without recoil due to the low caliber. Combine its light weight, firepower, lack of recoil and laser sight, and you have what a spokesman for the manufacturer calls "a downright nasty gun."

In use by an estimated 300 police departments and military organizations, the American 180 can drill through wood, concrete

Using laser "target designators," soldiers in the field could give precise range and location data to fighter jets, artillery, and laser-guided weapons. (Credit: Hughes Aircraft)

X-Ray Lasers

An x-ray laser system with the potential to "tip the battle in favor of the defense for the first time in the history of nuclear warfare," has been developed by the U.S. Defense Department at Lawrence Livermore Laboratory.

Tested at an underground site in Nevada, the compact device is pumped by x-rays from a small nuclear blast. It operates in the x-ray wavelengths, delivering a powerful beam carrying trillions of watts in pulses lasting only billionths of a second.

The device looks like a car distributor cap stuck with a circle of needles. This rather mechanical pin cushion is described as "an x-ray laser ring." The needles are rods of solid, dense material circling the nuclear detonator. Each can move to track a target.

Unlike chemical lasers, the x-ray rings would destroy their targets with a strong "kick" delivered by the beam's shock wave.

The devices are small enough to be placed in orbit by the space shuttle. A sufficient number to protect the entire U.S. from nuclear missile attack could be put in play in a single space-shuttle trip. In addition, they could be placed on boosters for launch during a crisis.

U.S. officials connected with laser weapons development believe this development will considerably enlarge the number of options available for defending against nuclear attack. They have stated that 20 to 30 x-ray ring battle stations might be enough to protect us against a massive Soviet attack.

and even metal car doors with its rapid rate of fire. The laser sight projects a beam of ruby light that spreads no larger than a quarter at 50 feet, and to only three inches at 600 feet. Trained on a distant target, it practically ensures a hit. In field tests conducted by the U.S. Army, the number of hits achieved on stationary and moving targets were termed "unbelievable." They were 50 percent higher than with any other rifle ever tested.

An actual event in Fort Lauderdale, Florida, in which police used the weapon to stop two men who had robbed a grocery store, illustrates both its lethal penetrating power and a side benefit. During the chase, one robber turned and fired at a policeman carrying the 180 rifle. The officer put the laser dot at the robber's car door and fired. The bullets battered through the door and a fiberglass seat. In a brief burst of fire, the officer put 47 slugs into the car. Nine lodged in the robber's heart. None left the car. By contrast, another officer fired a shotgun at the fleeing criminals, and its pellets scattered. Because of this accuracy and control compared to other weapons, police believe

Here, a technician uses an automatic laser inspection system to insure that the Army's laser target designator will work once it lands in the field.

the 180 is less dangerous to innocent bystanders. In addition, it has had another side benefit: criminals who have heard of its reputation tend to drop their weapons and surrender when the red laser dot appears on them.

Although laser weapons are the flashiest military use of the technology of coherent light, they have many other uses of interest to the Armed Forces. The Navy is experimenting with blue-green lasers, which penetrate water, even dark ocean water, to determine whether they might aid in submarine communications and detection (see Chapter 6.) The Army uses lasers as range finders that shoot an invisible infrared tape measure to a target miles away, allowing accuracy never before possible with a variety of conventional weapons.

The future seems to hold an even larger role for laser weapons. Lasers pumped by nuclear devices, electron beams, chemical means and even other lasers, have extended their power and range to levels where there is no longer any doubt that they can perform as lethal weapons. While their use in space—where a natural supercold vacuum exists, making powerful lasers easier to construct and operate— is virtually assured, current research suggests they will play a major role in ground combat as well.

Among the latest developments in laser weapons technology is a device that produces a shock force rather than a heat blast and is tunable to many frequencies not available with conventional lasers. Another high-energy system designed by United Technologies Research Center uses a doughnut-shaped radiation pattern to guide its primary laser beam through the atmosphere without thermal blooming and turbulence effects. These two problems have been one of the major obstacles to developing a truly effective ground-based laser system. The United Technologies system "has a potential for reaching out to very extended ranges under conditions that would completely obviate (stop) normal laser operations," claims one of the company's scientists.

Perhaps, however, it is not too much to hope that Senator Wallop's attitude toward laser weapons turns out to be prophetic: "It is a benign weapon . . . fitted only for destroying instruments of mass destruction . . . in space. It can only save lives."

This is an artist's conception of the newest item being tested for possible use as a space weapon, the X-ray laser. (Credit: Betty L. Wright)

Chapter 5

How the Laser Works

All lasers work on the same basic principles. The atoms or molecules of some material are "excited" to amplify the intensity of light on a subatomic level by stimulating emission of coherent photons. The material may be a solid (such as a ruby), gas (such as carbon dioxide) or liquid (such as dyes). As one laser specialist notes, however, "it seems like only a mild exaggeration to say that if a particular material has not yet given laser action, we simply have not 'hit' it hard enough." Laser action has been produced by several odd substances, including human breath and Scotch whiskey vapors. The Russians have used vodka, and at Stanford, scientists produced "the first edible laser" with a block of gelatin (jello) containing a fluorescent dye.

Whatever the material used, it is arranged in a long column or rod with mirrored ends. One mirror is only partly silvered or polished and reflects less light than the other. The space between the facing mirrors varies from less than a millimeter in semiconductor lasers, to several meters in gas lasers.

PUMPING

The atoms or molecules of the material used are excited by pumping in energy from an outside source. The electrons of atoms travel around their nucleus in fixed orbits. Lower orbits have less energy than higher ones. When energy of the right frequency strikes electrons they are kicked to a higher orbit or excited state. Lasers can be pumped by powerful flashlamps, other lasers and even atomic explosions. When an excited atom is smacked by photons, stimulated emission results—the atom refunds the pumping energy *and* the pumping photon. The electrons give this energy back in a special way—as coherent photons of a single frequency, a single color, their waves locked in step. This produces the laser's powerful beam.

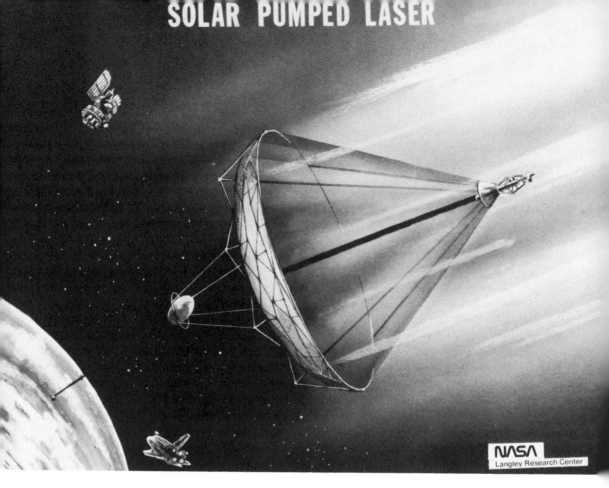

Lasers can be "pumped" with powerful flash lamps, electric discharges, or even sunlight or nuclear reactions. The artist's concept of a solar pumped laser shown here illustrates how this device would reflect the sun's rays from a disk to a gas-filled tube, exciting the atoms to create a high-intensity laser beam. Solar pumped lasers could be used to convert sunlight into electricity for use on earth. (Credit: NASA)

OSCILLATION

Oscillation is the process of amplifying energy to make it stronger and more useful. In a laser this is accomplished by forcing the stimulated photons of light to bounce back and forth between the mirrors in the laser cavity or resonator. (See illustration, page 69.) Photons not zipping back and forth parallel to the mirrors escape out the sides of the cavity walls. As the photons bounce to and fro, they stimulate additional coherent light. When it reaches a certain intensity, the coherent beam emerges from the partially mirrored end of the laser.

FOUR TYPES OF LASERS

There are four main types of lasers: solid-state, gas, liquid and semi-conductor.

Solid-state lasers are made from a solid rod of crystalline material such as ruby. Impurities in these crystals, such as chromium in ruby, have atoms that can be stimulated to emit coherent radiation by strong white light. Another laser of this type is made from yttrium aluminum garnet doped with neodymium. Doping is the process of adding an impurity necessary for lasing action. The full name of these garnet lasers is quite a mouthful, so scientists refer to them by the abbreviation YAG.

In a solid-state laser the crystal rod is pumped by a high-intensity flashlamp wrapped around it in a spiral. The lamps most often used are filled with the gas xenon and charged by electricity. The ends of the crystal rods themselves are polished to serve as mirrors. The rod of ruby or other crystal serves as the resonant chamber that amplifies the light.

Gas lasers resemble an ordinary neon tube, the sort used to make signs. Gas lasers include the powerful, continuous-beam carbon dioxide (CO_2) units which emit in the infrared wavelength; argon lasers which give blue-green light useful in surgery; helium-neon lasers; and those made from krypton, neon and other gases.

Ali Javan invented the first gas laser (helium-neon) in 1961 (See Chapter 6). The light from gas lasers is even more coherent than that from solid-state devices. If we compare light to sound, then ordinary light is a mixed din of noise such as you might hear at a large party, ruby laser light is a well-trained chorus and gas laser light is a pure, sharp siren blast.

Gas lasers are sometimes called "atomic lasers" because they rely on colliding atoms transferring their energy to create lasing action. In the helium-neon laser, for instance, a mixture of the two gases are placed in a quartz tube. Ten times more helium than neon is used. Special reflecting surfaces—the basic laser trick—bounce photons back and forth to stimulate emission. Pumped by an electrical discharge rather than flashlamps, the helium atoms climb to one of two energy states. These excited helium atoms smack into the ground-state neon atoms, transferring their energy to them. This shoves the neon atoms to a higher energy level. This energy transfer leads to a population inversion—more atoms at an excited state than at the ground state. At this point, stimulating them with energy of the proper frequency knocks them back to a lower energy level, forcing them to emit coherent photons. These zip back and forth between the mirrors until lasing action occurs.

Later gas lasers used different mixtures of these gases and various tricks with the mirrors and pumping techniques to achieve lasing action on several levels. This opened the way to producing additional wavelengths of laser light. Still other gas lasers split oxygen molecules (O_2) into two separate atoms using excited neon or argon atoms. Resulting energy transfers produce the population inversion necessary for lasing action.

Gas ion lasers.   A third way gas lasers can operate is with only one of the noble gases (argon, neon, krypton, xenon, helium). Pumped by a radio frequency discharge, these lasers work because the electrons of the discharge collide with the atoms of the gas used. They produce laser light over hundreds of wavelengths. New krypton lasers that are used to pump dye-liquid lasers are giving scientists a way to explore chemical reactions with very short pulses of light. (See Lasers in Scientific Research, and Lasers in the Future).

Carbon dioxide lasers.   Invented by C.K.N. Patel of Bell Labs, the powerful carbon dioxide laser is the workhorse of the laser field, used when relatively efficient power is needed. The CO_2 laser obtains efficiencies of 10 percent—it turns out 10 percent as much energy as is needed to make its gases lase. (Ten percent may not sound like much, but most lasers are much, much less efficient. The appalling lack of efficiency in lasers is one of their major limitations.)

Molecular excitation produces its laser action. Contained in a long gas tube, the CO_2 is pumped by an electric discharge. Its energy excites nitrogen molecules, and these, in turn, make the single carbon atom vibrate back and forth between the oxygen (O_2) atoms on each side of it. When the excited CO_2 molecule drops to a lower energy state, it emits a photon. These strike other molecules, stimulating the release of still more photons, which multiply and gain coherence in the resonator.

Liquid lasers, also called Raman lasers, were discovered by accident at Hughes Labs. In 1962, researchers there were using a "cell" of liquid nitrobenzene to block the mirror shutter of a ruby laser (a process called Q-spoiling; *see* A New Light). The scientists, Eric Woodbury and others, noticed that about 10 percent of the ruby laser's light seemed to be missing. They found that the dye cell absorbed the

TYPICAL GAS LASER CONFIGURATION

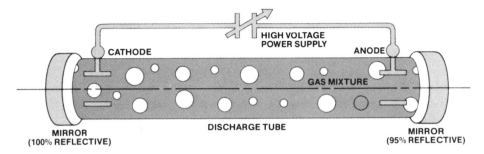

This diagram shows how a CO_2 laser operates. A mixture of CO_2, helium, and neon in a glass tube is the lasing medium. A high voltage power supply excites the atoms of the gas mixture. Photons reflect back and forth between the mirrors in the optical resonator chamber. One mirror is partly transparent and allows some of the light to emerge as a laser beam. (Credit: Coherent Radiation)

light and re-emitted it as coherent light at a different wavelength. This occurred because of what scientists call "the Raman effect." In the ordinary Raman effect, light is scattered from molecules. This outgoing, scattered light is at a lower frequency than the light that went in. Molecular vibrations within the liquid account for the energy difference.

The Hughes scientists, by stimulating this effect in the liquid with a laser, obtained a lower-frequency coherent beam.

Organic dye lasers.   In 1966, IBM scientists Peter Sorokin and John Lankard found the key to obtaining lasing action from organic dye solutions. These lasers are special because they can be "tuned" over a wide range of frequencies . . . hundreds of angstroms. Also, a dye laser's beam can be concentrated into an extremely narrow band by using optics such as diffraction gratings to disperse the light for color selectivity.

The lasers are tuned much as a radio is, by turning a dial—in this case, rotating the diffraction grating. The dyes used are the kind that color fabrics. They have a high fluorescence efficiency (their atoms give off photons of light easily). A ruby laser was required to pump these lasers. A very short pulse lasting only nanoseconds (billionths of a second) is used because otherwise these dyes decay into a state in which lasing action is impossible. Later, the IBM scientists devised a simplified method of pumping the dyes with flashlamps.

Dye lasers give frequencies ranging from the infrared through the visible region to ultraviolet. Their combination of brightness and tunability has made them extremely useful in chemical and physical science research. They can be flashed in ultrashort pulses lasting only billionths of a second and less, enabling them to perform measurements on a scale never before possible. (For applications of organic dye lasers, *see* Lasers in Scientific Research.)

Semiconductor (injection) lasers   A semiconductor is a substance that conducts electricity better than nonconductors such as rubber, but not as well as copper. Germanium and silicon are semiconductors used to make transistors. The memory chips in calculators and electronic games are made from silicon. Gallium arsenide is the semiconductor first used in lasers.

To make a gallium arsenide semiconductor work as a laser or transistor, it is "doped" with two impurities, zinc and tellurium. Sandwiched between them, the gallium arsenide is a very thin layer. The entire semiconductor crystal is no longer than a speck of dust. Lasers made from them are dwarfed by a grain of salt.

The semiconductor sandwich has three regions. The tellurium region has an excess of electrons (more than the arsenic it replaces) and is called the N-region (N represents *negative*). The zinc region has fewer electrons than the gallium it replaces. This creates electron "holes." This is the P-region (P means *positive*). Between them is the pure gallium arsenide buffer zone, which is called the junction. Injected with a powerful jolt of electricity, the crystal produces laser action in the 1/10,000-inch-wide junction.

When the current flows through the crystal, the extra electrons in the N-region cross the junction to the holes in the P-region. As the electrons drop into the holes, they change energy levels, giving off a photon as they do so. If the two sides of the crystal have been polished, the photons bounce back and forth and laser action results.

Modern, state-of-the-art semiconductor lasers are already being used for fiber-optic communications by Bell Telephone.

CHEMICAL, RARE GAS, ELECTRON AND X-RAY LASERS

Chemical lasers use the energy created by chemical reactions to produce laser light. A deuterium fluoride, continuous-wave chemical

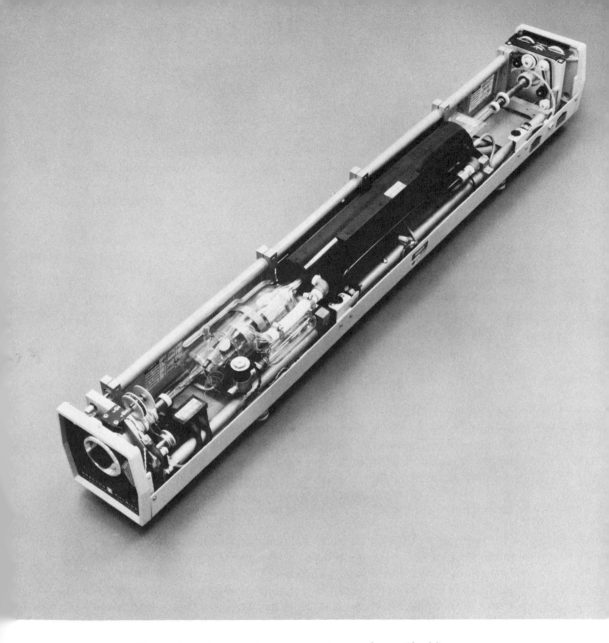

The inside view of a 5-watt argon-ion gas laser. The blue-green light produced by these devices has proved extremely useful in surgery and many other applications. (Credit: Spectra Physics)

laser is considered a leading candidate for air and missile defense laser systems at present. High-intensity light pumped into chemical lasers breaks a molecular bond. Free atoms split from the molecule by the light are in an excited state and will produce laser light.

Rare gas or excimer lasers use a molecule that exists only when one or both of its elements are in an excited state. The word excimer is derived from "dimer." A dimer is a compound joining two molecules of a simpler compound. An excimer is an excited dimer.

Krypton/fluorine rare gas halide excimer lasers involve a complicated process. Basically, excited krypton combines with fluorine to form a salt, giving off laser radiation in the process. Unexcited krypton atoms and fluorine repel each other.

Free electron and x-ray lasers are among the newest items in laser technology. The free electron laser was developed at Stanford as an outgrowth of the university's work in particle acceleration. They require huge machines and high power inputs. Their prime characteristic is that they can be tuned.

A MARTIAN LASER

Natural lasers.   Astronomers have discovered both natural masers and lasers. Radio astronomy reveals that the Orion nebula, where stars are being formed, is a natural maser pumped by infrared light from nearby collapsing gas clouds. Several red giant stars also broadcast maser radiation.

Perhaps most surprising to scientists, however, is that the upper atmosphere of the planet Mars emits naturally created infrared laser light.

Located 50 miles above the planet's surface, the light seems to come from carbon dioxide molecules scattered throughout the Martian atmosphere. Sunlight excites the molecules to give off photons, and these photons in turn bump other excited molecules, stimulating emission of coherent laser light. This produces infrared "light" one billion times stronger than the cold Martian atmosphere (−250 degrees Fahrenheit) should create.

The natural laser was observed with an infrared detector at Kitt Peak Observatory by physicist Michael Mumma and associates from the Goddard Space Flight Center.

Naturally occurring lasers and masers may enable humans to learn new ways to produce such devices on earth.

Chapter 6

Listen to the Light:
Lasers and Communications

THROUGH THE LASER CRYSTAL BALL—
A DAY IN THE LIFE

At 7 AM, Bill Taylor's computer-run home beams a laser-pulsed message to his alarm clock, which wakes him. Bill lingers in bed for a time, but every two minutes the clock nags him . . . "It is now four minutes past your alarm time . . . It is now six minutes past your alarm time . . ."

Sliding out of bed, he pauses for an instant on the edge and taps a button on the bedstand console. "News and feature sections of the *New York Times*, business section of the *Wall Street Journal*, local section of the *Bloomsburg Morning Press*," he instructs the computer. He hears the printer begin to purr in the living room—the laser-etched copy will be waiting for him before he completes his shower.

Taylor scans the papers while eating breakfast later. He particularly enjoys glancing at the local news from his hometown in Pennsylvania. Flipping the computer sheets, he thinks it may be time to start ordering the papers on one of the monitor screens rather than on paper. The habits of a lifetime are hard to break, however, and he enjoys holding the printed words in his hands. Besides, in the morning he likes to skim the headlines, mentally noting articles to read more fully later.

After breakfast, Taylor wanders into his home office early. Looking over his day's agenda on the desk monitor, he finds it difficult not to recall—as he still does nearly every morning—that just a few years ago, he would be locked in traffic for the next hour on the way to his office. Laser communications piped through fiber-optic cables made it possible for him and millions of other workers, from company presidents to secretaries, to work at home. The savings in energy and time made it practical. Taylor, who had always hated the time and cost of commuting, the aggravation of traffic and the lousy lunchtime meals, gratefully rapped his computer with his knuckles. *I'd say 'knock on wood' if any were around,* he thinks.

When he completes his day's work, he quits. Although he remains available if needed, his flextime schedule allows him to start and stop as the work itself demands. Retiring to the living room, he flops in his favorite chair and turns on the wall screen TV with the remote. He wants to watch a football game before the kids finish school. With more than 500 channels catering to every imaginable taste, there is always an argument as to who will watch what on the giant, laser projection screen.

This futuristic vision may sound fanciful—it isn't. Although it's science fiction now (just barely), it's the kind that comes true. Communications play a pervasive role in today's work and leisure activities; lasers will play a pervasive role in tomorrow's communications. In fact, your telephone calls may already be zipping through a fiber-optic cable carrying 44.7 million bits of information each second—but we're ahead of our story. To understand why laser light communication is needed in today's world, and may transform tomorrow's, we must again look back to yesterday.

THE GLOW OF LIFE

In his amusing Public Broadcasting Service television series and book, *Connections*, British writer James Burke traces the modern telecommunications revolution back to one night in 1675. On that night, Burke notes, the French astronomer M. Jean Picard, walking home from a stint at the Paris Observatory and "happily swinging his barometer," realized that the instrument had begun to glow. Interest in this "glow of life," as it was called, led, bit by bit, to exploration of the seemingly magic force that crackled and sparked like lightning and made certain substances glow when they were agitated by friction, as was the mercury in Picard's swinging barometer. Since one of the substances that could be made to glow in this fashion was amber, the force was called electricity, from the Latin word for "amber," *electrum*. The Latin word, appropriately enough, derived from the Greek word ēlectron, which meant gleaming, shining or brilliant— which the new force certainly was.

Despite a long series of experiments by a host of researchers which followed, it was not until more than a hundred years later that scientists began to understand this force well enough to use it. In 1820, the Danish professor H.C. Oersted ended a lecture with an experiment designed to show that no connection existed between electricity and magnetism. He placed a compass, with its magnetic needle, close to a wire, intending to show that sending an electric current through the wire would not affect the compass. To Oersted's surprise, the needle moved as soon as he switched on the current. Oersted rightly concluded that the electric current in the wire must have a magnetic field as well.

Soon, discovery followed discovery. Scientists made magnets using electricity (William Sturgeon) and electricity from spinning magnets (Michael Faraday). The German scientist Hermann von Helmholtz, who was also a pianist, noticed that switching an electromagnet on and off could make the arms of a tuning fork vibrate, producing a sound. One of Faraday's admirers, the Scotsman James Clerk Maxwell, used brilliant mathematics to demonstrate that electromagnetic forces had some of the properties of waves, and suggested (long before the knowledge would be practically useful) that light, too, was a form of electromagnetic energy.

In 1875, Alexander Graham Bell put the work of many of these men and others together and invented the telephone (and was also the first person to communicate sound with light—see Time Capsule, page 93).

Although the modern communications system had begun, it still had one particularly severe limitation: It required wires for transmission of electrical signals. Telegraph wires (strung for the first time in 1844), trans-Atlantic cables (1866) and Bell's telephone all required wires.

In 1885, however, the German physicist Heinrich Hertz discovered radio waves. Hertz found that these waves, much longer than those of light, could be transmitted across great distances and easily penetrated matter, all without wires. With the Italian engineer Guglielmo Marconi, Hertz, in 1896, sent a wireless signal to a receiver nine miles away. On December 12, 1901, he transmitted a radio wave across the Atlantic ocean to the United States. While at first only dots and dashes could be sent on radio waves as the current was switched on and off, the American scientist Reginald Fessenden invented a generator that could "modulate" the current. This meant the waves could be patterned. If the pattern mimicked the vibrations produced by sound, they could carry a human voice. By 1906, voices and music were carried on waves broadcast through the air.

This short jaunt backwards in time has concentrated on the technology of communications. From the very beginning, researchers

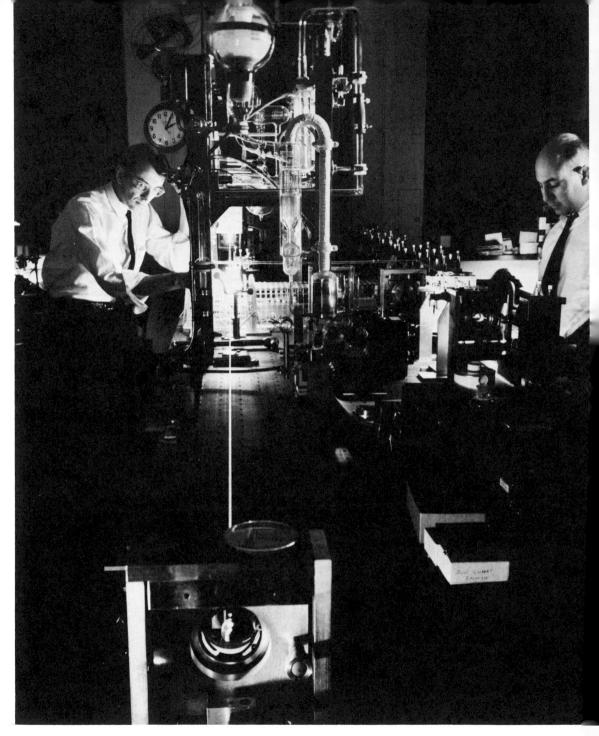

Several hundred gas and solid-state lasers have been discovered at Bell Laboratories. This is the first gas laser to generate a continuous beam of visible light (1962). (Credit: Bell Labs)

have tried to make electromagnetic waves carry more and more messages on more and more frequencies.

The frequency of electromagnetic waves, as you may recall, is related to wavelength; the higher the frequency, the shorter the wavelength. Also, as frequencies increase and wavelengths decrease, the amount of information a frequency can transmit is correspondingly raised. The radio frequency range of the electromagnetic spectrum (see diagram, page 82) includes very low-frequency waves that stretch for thousands of kilometers (a kilometer is slightly more than half a mile) to the microwaves, which are measured in centimeters (not quite four-tenths of an inch). The longest of these waves have frequencies of 10 cycles per second (they go up and down 10 times a second), while the shorter wavelengths have frequencies of billions of cycles per second.

The vast range of frequencies and wavelengths in the electromagnetic spectrum accounts for the use of various units of measurement to talk about them. Essentially, however, for our purposes here, it is necessary to remember that frequencies are measured in cycles per second (which are also called hertz). These cycles per second are usually preceded by a metric prefix such as kilo (1,000), mega (1,000,000) or giga (1,000,000,000). Thus, thousands per second frequencies would be measured in kilohertz (kHz.); frequencies of millions of cycles per second are measured in megahertz (mHz.); and very high frequencies of billions of cycles per second are referred to as gigahertz (gHz.).

The more cycles per second a frequency has, the more room it has to carry information. If a voice signal is converted to electricity, each message requires from 200 to 3,500 cycles per second to cover the range of sound wave vibrations that human ears can hear. The on-off messages of Morse code can be carried on longer, lower-frequency waves. But it is not difficult to see that the higher frequency range, with its spread of billions of cycles per second, can handle a great many more messages of 200 to 3,500 cycles per second each.

THE BABBLING POINT

When the first telephone systems were built, each wire carried only one message. In today's world, with each telephone line carrying hundreds of calls, and each coaxial cable transmitting thousands of calls an hour, it is still impossible to get a call through to some cities at

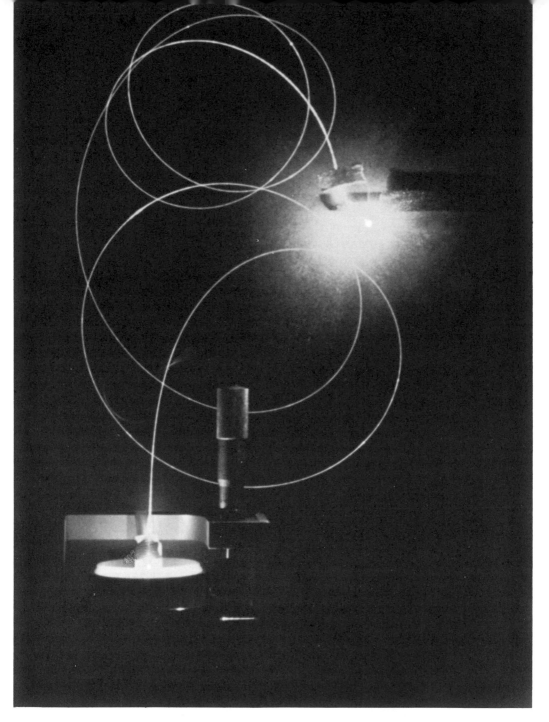

This laser-pumped, low-loss optical fiber is already providing increases in message-carrying capacity for the Bell System. Less than the diameter of a hair, these optical fibers can
carry hundreds of messages on a single strand. (Credit: Bell
Labs)

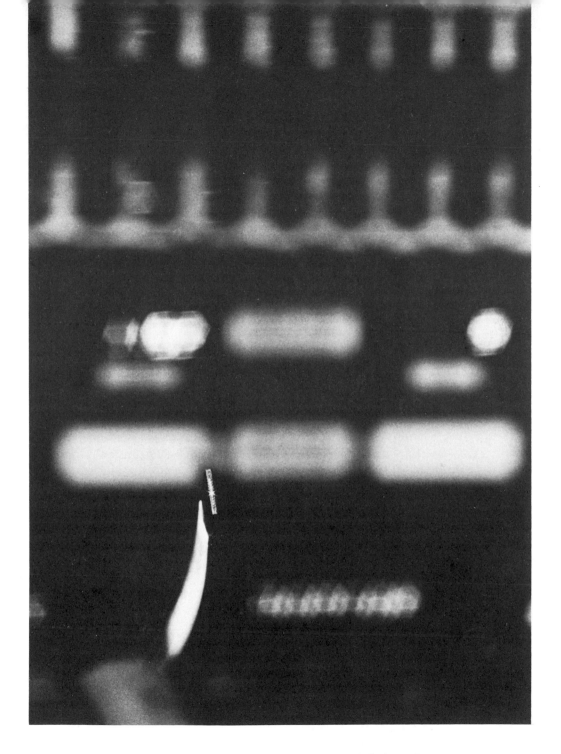

A sparkle of laser light illuminates the end of a lightguide in a ribbon of glass fibers. This tiny ribbon can replace dozens of thick copper phone wires. (Credit: Bell Labs)

certain business hours. In fact, the vast increase in the use of tele-communications in recent years has crowded most available frequen-cies to the babbling point. Teleconferences, Telex systems, comput-ers, radio and television worldwide, satellites and telephones every-where have jammed communications circuits to near bursting. Even the energy crisis has added to the burden communications must carry in the daily conduct of business internationally; business is con-ducted more efficiently, quickly and inexpensively via computer, Telex and telephone.

CHARGE OF THE LIGHT BRIGADE

So, the search for additional higher frequencies capable of carrying information has continued. Radio broadcasts currently extend from 10,000 cycles per second for ship-to-shore maritime communications to the 200 megacycle (200 million cycles per second) of Ultra High Frequency (UHF on your TV dial) television channels. Some special, experimental uses drift up even higher on the spectrum. Until the invention of the laser, however, the frequencies above the radio waves offered little hope of being used for communications. From infrared to visible light, this band width of high-frequency electromagnetic waves appeared tantalizing, yet impossible to deal with using vac-uum tubes, transistors and conventional radio apparatus.

Though we may think of light waves as very different from radio waves, their only real difference is that they vibrate (in electronics terminology, oscillate) up and down much faster. In other words, their frequency is higher. The waves of light range from a thousand to a million times higher in frequency than microwaves. This means that they have the potential of providing up to a million times more space to pack with messages than the entire range of radio frequencies now offers. The visible light band alone could accommodate 80,000,000 television channels or 800,000,000 telephone calls. Think about try-ing to decide what to watch on TV with even a tiny fraction of that number of stations in service! And that number of telephone calls would allow every man, woman and child in China to make a simul-taneous call without tying up the lines.

Not very surprisingly, much of the research in using laser light as a communications tool has been done at Bell Telephone research laboratories. Dr. Charles Townes began his pioneering work with quantum devices there, and Dr. Arthur Schawlow, who collaborated with Townes on the concept-breaking description of the laser in the late 1950s, never received any money other than salary because Bell holds the patent on the device he proposed. As Dr. John Asimus, laser expert at Maxwell Labs in California, points out, "You can trace almost everything that's happened in the laser field back to either Bell Labs or the defense department, although there are some exceptions."

THE CONTINUOUS GAS LASER

One of these major laser developments, the incredibly useful gas laser, resulted from early work at Bell. Describing this event in the Bell publication, *Laser: Challenge to Communications Science,* Manfred Brotherton wrote:

"On a December afternoon in 1960 at Bell Laboratories in Murray Hill, New Jersey, three scientists had been experimenting in a darkened room with a long glass tube that glowed like a neon sign. The 40-inch tube was connected to an assortment of mirrors, lenses, and electronic equipment that would have appeared strange indeed to an uninformed passerby."

In this mad-scientist-style lab setting, Brotherton notes, many previous experiments had taken place . . . and failed. The laser project had been started many months earlier. Bell researcher Ali Javan figured out a way that it might be possible to generate a continuous laser beam. This particularly interested the Bell scientists because a continuous beam, as opposed to the pulsed flashes of the only other lasers then available, would be necessary for use in communications.

Javan's idea evolved from his knowledge of the nature of materials that modern spectroscopy had provided.

"Tucked away in the storehouse of data" about the behavior of matter, explains Brotherton, "was a fact about certain gases that caught Javan's imagination." When these gases are mixed in pairs and one gas is excited, its higher-level energy becomes trapped. At the same time, the mixed atoms of the gases clash and collide in ceaseless, violent motion. When an atom of the first gas, with its locked-in energy, collides with an atom of the second, the trapped energy is transferred from one to the other. The atom of the second gas immediately releases this energy as radiation. Javan saw the possibility of harnessing this energy to charge a laser.

Javan thought it might work like this: a mixture of the two gases would be enclosed in a long glass tube. Then, radio frequency energy would be pumped into the tube, exciting the atoms of one gas. This

WAVELENGTH in meters	ELECTROMAGNETIC SPECTRUM			FREQUENCY cycles/second
10^{-14} to 10^{-17}	COSMIC RAYS	Invisible, super-short waves from outer space		-10^{25} -10^{24} -10^{23} -10^{22}
10^{-11} to 10^{-13}	GAMMA RAYS	Radioactivity released by nuclear reaction		-10^{21} -10^{20}
10^{-9} to 10^{-11}	X-RAYS	Medical X-rays X-rays used in diffraction studies		-10^{19} -10^{18}
10^{-7} to 10^{-9}	ULTRAVIOLET RAYS	Sunburn Fluorescent band		-10^{17} -10^{16} -10^{15}
10^{-6} to 10^{-7}	VISIBLE LIGHT	Blue Green Yellow Red		-10^{14}
10^{-3} to 10^{-6}	INFRARED	Radiant heat		-10^{13} -10^{12}
10^{-3} to 10^{-6}	EHF (Extra High Frequency)	RADIO WAVES	Radar Microwaves	-10^{11}
10^{-1} to 10^{-2} (1–10 cm)	SHF (Super High Frequency)		CB and police radio	-10^{10}
1 to 10^{-1} (10–100 cm)	UHF (Ultra High Frequency)			-10^{9}
1 to 10	VHF (Very High Frequency)		TV	-10^{8}
10 to 100	HF (High Frequency)		Commercial radio	-10^{7}
100 to 1,000	MF (Medium Frequency)			-10^{6}
10^{3} to 10^{4} (1–10 km)	LF (Low Frequency)			-10^{5}
10^{4} to 10^{5} (< 10 km)	VLF (Very Low Frequency)		Ship-to-shore radio	-10^{4} (10,000)
	ELECTRIC POWER and LIGHT			-10^{3} (1,000) -10^{2} (100) -10

gas would absorb the energy, then transfer it to the other through collisions. The second gas would shed this excess energy as light radiation. Racing back and forth between the mirrored ends of the tube, the photon energy would gain coherence and power until it burst forth as a continuous beam of laser light.

Unfortunately, in trial after trial, the all-important mirrors "cracked in fine lines like old china," requiring new equipment. On that winter day in 1960, though, chance, in the form scientists sometimes call "serendipity" (making fortunate and unexpected discoveries by accident), played a role. Once again the tube's mirrors cracked. The two Bell researchers working on the device, optical specialist Donald R. Herriott and physicist William R. Bennett, Jr., paused to discuss what had gone wrong. As they did so, Herriott almost absentmindedly continued fiddling with the knob adjustment of a mirror. Suddenly, he saw a peak in the oscilloscope trace and realized the long-sought lasing action had been achieved. All that had been needed was a slight adjustment of the mirror.

"MR. WATSON, COME HERE, I WANT YOU"

By the next day, the laser operated at top efficiency, producing a continuous, invisible infrared beam. While demonstrating the laser to colleagues, they noticed that whenever they spoke near the carefully adjusted mirrors, the sound vibrations from their voices produced similar vibration in the laser's output. They attached an earphone to

In this chart of the known electromagnetic spectrum, wavelengths are measured in meters, frequencies in cycles per second (also called Hertz). It ranges from the 60 cycles per second waves put out by power and light companies, which are 4,800 kilometers (3,000 miles) long, to the ultra-tiny cosmic rays. There is no theoretical limit to the spectrum on either end. A note about scientific notation may help you understand the chart. In science and engineering, it is often necessary to deal with astronomically huge or very, very tiny numbers. To express these numbers, scientists use a base number (such as 10 in this chart) raised to a certain power (exponent). The power or exponent expressed under the base number preceded by a plus or minus sign indicates that the number is to be multiplied by itself (10^3 equals $10 \times 10 \times 10$)

the oscilloscope that monitored the laser's beam . . . and heard their own voices.

That day, when Dr. Javan (who was away in New York) called the lab, Herriott and Bennet took the opportunity to reproduce a famous moment in the history of communications science with a modern twist. Holding the earphone—which was still attached to the laser monitor—near the telephone mouthpiece, they shouted at the laser mirror and talked to Javan through this means. "Mr. Watson, come here, I want you," they said, repeating Alexander Graham Bell's historic first words over his telephone.

Since communication in any given frequency range cannot take place until the tools are available, development of a continuously operating laser was one step toward advancing into the vast spread of frequencies above the microwave region. But it was only the first step necessary in a long hike, for light, even laser light, has serious drawbacks if used to send messages over any distance. All light, even the intense, coherent light of a laser, is absorbed, scattered or reflected by fog, smoke, rain, snow and the atmosphere itself. A whole new communications technology proved necessary to take advantage of light's vast frequency range. Without it, a whiff of smoke, a stray cloud or falling snowflakes could stop the charge of the light brigade in its tracks.

THE LIGHT GUIDE: FIBER OPTICS

When a technical need exists, and no fundamental law of science prevents its development, it's a good bet that engineers, inventors and scientists will find a solution to the problem. Fiber optics married to lasers have solved many of the problems involved in communicating with light.

Fiber optics are extremely clear glass tubes half the diameter of a human hair. Originally developed for use in medical instruments called endoscopes in the 1950s, fiber optics are made from silicon, the same material that is used to make the memory chips in calculators and microprocessors. The special manufacturing process called Modified Chemical Vapor Deposition (MCVD) developed at Bell gives fiber optics a number of unusual qualities.

In this process, bubble-thin layers of silica particles (glass) coat the inside of a particularly pure quartz tube when it is drawn to and fro over a torch. These silica coatings have refraction properties that bend stray light back to the center of the tube, preserving the clarity of the ray traveling through it.

After a solid glass tube called a preform is made using the MCVD process, it is softened in a furnace, then stretched into a thin, long fiber. This almost flawless fiber has none of the impurities, pits or cracks that cause ordinary glass to break. With its flaws limited to those no larger than a micron (smaller than a bacterial virus) and occurring less than every kilometer, optical fibers are twice as strong as steel. Each fiber thread can take more than 600,000 pounds of pulling force per square inch.

The fibers are so fine that 144 of them can be comfortably contained in a finger-thick cable. Each pair of fibers is capable of carrying 672 simultaneous phone conversations transmitted on laser light pulses. The 144-fiber cable can handle 50,000 calls—which would require a four-and-a-half-inch thick cable of 4,050 pairs of conventional copper wires.

As a sidelight, it is interesting to note that telephone conversations on both the old and new systems use what Bell calls the "time-division multiplexing" system. This sytem chops your voice into little bits and fills the minute pauses in your speech with other conversations. The tiny spaces between your spoken letters, syllables and words, are large enough to fill with other people's talk. In part, this is why we sometimes hear other conversations on our lines.

THE CRYSTAL BOX: SEMICONDUCTOR LASERS

The final step in producing communication via light came in the early 1970s: solid-state semiconductor lasers. Made from a crystal of laboratory-created gallium arsenide substrate, these lasers are so small that Bell scientists joke about the danger of inhaling them accidentally. Smaller than pinheads, they generate a narrow beam of light and are easily joined with an optical fiber five thousandths of an inch in diameter. (As can be seen in the photo, these lasers are so small that a single grain of salt dwarfs them.) One reason these lasers can be made so tiny is that they do not require the mirrors and other paraphernalia of ordinary lasers. The crystals used to make them have facets (the multisided faces easily seen on the familiar diamond crystal) which can be polished to bounce light back and forth inside.

Furthermore, these solid-state lasers have been perfected to the point where they have average projected lifetimes of one million hours—which equals 100 years. "These lasers now last so long that it's hard to determine their expected lifetime in a system application,"

Much smaller than the grain of salt to its right, this solid-state laser (small rectangular atop block) may hasten the day you talk over a beam of light. (Credit: Bell Labs)

says Barney DeLoach, head of the Bell Labs light wave sources department, which fabricated and tested them. Because of this factor, they meet Bell System's reliability standards for a light wave communication system.

PUTTING IT TOGETHER

Combined with fiber optics, semiconductor lasers have made light wave communications possible. Dr. Walter P. Sigmund of the American Optical Corporation, quoted in the *Techno-Peasant Survival Manual*, explains the combination this way: "A light beam travels through an optical fiber much like a bullet ricocheting down a steel pipe. The beam caroms through the fiber's core, trapped there by the cladding (the delicately applied silica lining). The cladding . . . provides a mirror effect, turning the light back into the core. This creates what is known as total internal reflection. It is so perfect that you can have millions of such reflections through many kilometers of fiber and still have a light beam emerge largely undimmed."

Messages zap through the fiber in tiny, on-off blips of laser light. The light is modulated to carry sensible information in much the same way as radio waves are. Sound waves are turned into electrical waves in the mouthpiece of the telephone. A sampler device measures the height (amplitude) of each wave 8,000 times per second. By giving each tiny sample a binary number (the same number code used in many computer languages), the system describes each sound electrically. All of this is done at the rate of 44.7 million bits of information a second, which is fast indeed, and even higher information loads may be possible. The binary coded light blips are then reformed into sound waves at the opposite end of the conversation.

TELL ME MORE . . . QUICKLY

None of this is waiting for the future. Light wave communication is a fact. Your own telephone calls may have already zipped through fiber-optic cables riding pulses of laser light. Putting together 20 years of research, Bell successfully tested optical communications systems in Atlanta in 1976, and in Chicago from 1977 through 1979. Customized installations utilizing fiber-optic cables and lasers linking phone company offices are in use elsewhere. The first commercial

---

Time Capsule: Alexander Graham Bell Invents the Photophone

Four years after he invented the telephone, Alexander Graham Bell demonstrated a new device he called the photophone. It was 1880, one hundred years before researchers would find a practical method for communicating with light.

Bell's device, which resembled a long ironing board with a speaking tube and lamp at one end and a lens/mirror combination at the other, did not send light wave messages more than a few feet. Yet Bell himself was fascinated by his invention and the idea of piggybacking a spoken message on a light beam. Although the difficulty of extending the range of his device resulted in its becoming a dusty, technological footnote to science, Bell rhapsodized about the invention:

"I have heard a ray of sun laugh, and cough, and sing!"

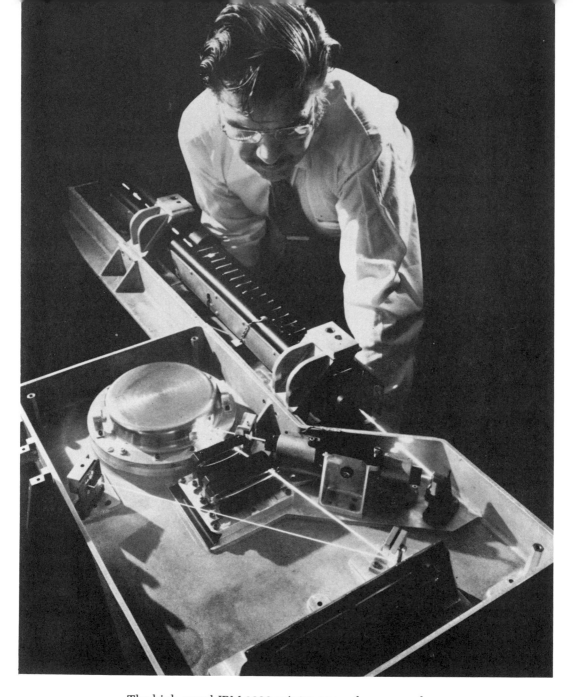

The high-speed IBM 3800 printer uses a low-power laser, shown here, to form character images on a rotating drum. As the drum turns, dry, ink-like powder that adheres only to the images is applied to the drum surface and transferred to paper, producing printed text. The 3800 operates at speeds up to 13,360 lines per minute to produce about 10,000 11-inch sheets of printed paper an hour. (Credit: IBM)

installation went into service in Atlanta, again linking the central offices that dispatch and receive calls from users in a given area.

LOOKING TO THE FUTURE

Further advances are expected, though. Among them are:

1. New ways to make fiber-optic glass even clearer so that it will carry light waves further than the four-mile limit now imposed before repeating the signal is necessary.

2. The manufacture of fibers from non-glass materials 1,000 times more transparent to some kinds of light than what is currently available.

3. A light-wave transatlantic cable that will carry hordes of information—from telephone conversations to television shows—to and fro across the ocean. Bell engineers are testing this concept in an "artificial ocean" to make sure that if they sink a cable two or three miles deep across 4,000 miles of salt water they "get it right the first time."

4. Fibers pumped with four or more lasers, each a different color and beaming separate information, vastly boosting the amount of information one fiber can transmit simultaneously.

LASER TELEVISION

Both in combination with fiber optics and by themselves, lasers are performing (quite literally in many cases) an ever-increasing number of communication functions. Laser light shows accompany rock concerts, classical orchestra performances and readings by, appropriately enough, the actor William Shatner, who portrayed Captain Kirk on *Star Trek.* Laser projections cast enormous images on buildings, slash light skywriting on clouds and rapidly carve words and pictures on printing plates. Laser images dance on Broadway stages; laser-pure colors create special effects on others. New artists working with lasers integrate a knowledge of light, color, form and shape to meld light with music, creating totally new, twentieth century experiences . . . or are they twenty-first century harbingers?

Large laser-projection television screens are among the more exciting of these coming attractions. Working laboratory models have already been built by several manufacturers. Large screen TV images projected by ordinary light sources tend to provide less than satisfactory viewing under many conditions. Dim images and grainy pictures reduce the impact of these sets, despite advances in rear-screen projection.

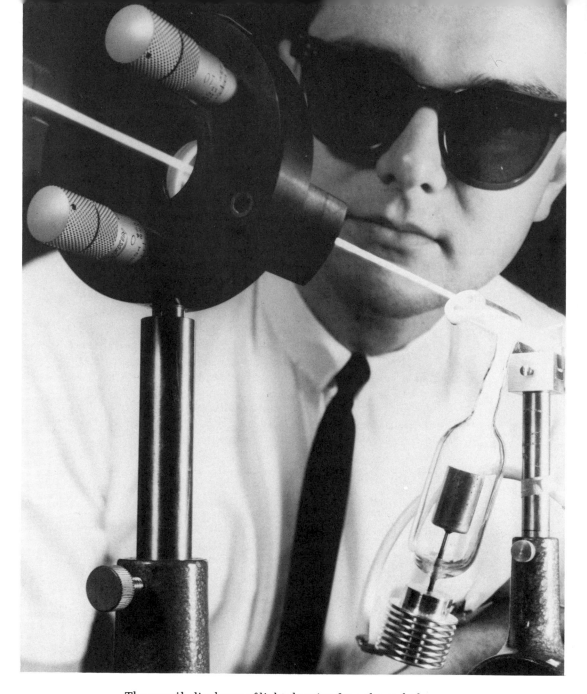

The pencil-slim beam of light shooting from the end of a gas laser (glass tube at right) through the mirror on the left can be tuned to produce more than 60 colors. This spans the entire visible portion of the spectrum and opens new possibilities for underwater investigations, space communications and color TV projection displays. (Credit: Hughes Aircraft)

Projected laser systems, though, might solve these problems. Current prototypes feature three lasers in red, green and blue. Ordinary picture tubes create images by shooting a stream of electrons at chemicals on the face of the tube, which phosphoresce (light up) to form a bright picture. Magnetically attracted back and forth over the face of the tube, the beam of charged electrons *scans,* creating a full screen picture in a fraction of a second. Laser light, unlike the charged electrons, does not react to magnetic attraction, so one problem in making a laser TV system was how to get a laser to scan. As early as 1966, Zenith found one answer. The company placed a fluid-filled chamber and electronic transducer in front of each laser. A transducer is a device that changes energy, such as that of the laser, from one form into another. By causing the fluid in front of each laser to vibrate, the transducer in the Zenith system makes the liquid act as a controllable lens. Through this means, by varying the output of the transducer, the lasers will scan.

Another possibility being seriously considered by several manufacturers is a laser projection system that will display holographic images suspended in space. Imagine the impact of watching a space launch in three-dimensional color from the comfort of your living room, with the space shuttle piggy-backed on its massive rocket, spewing flame and climbing toward your ceiling. Such full-bodied images might revitalize the beauty contest as a televised event, add new thrills to watching NFL linemen clash and quarterbacks pass, and give the arts (particularly dance and theatre) vibrant new life. Anyone who has seen a ballet both live and on television can appreciate the difference in impact; the same holds true for many events, from boxing matches to the Olympic Games.

Laser video disc display systems are already on the market, from several manufacturers. Since the picture information coded on these discs is read by a laser, they do not wear the way a record touched by a stylus does. They present sharp, bright images, but do have a drawback—the systems marketed to consumers do not have recording capabilities. Therefore, conventional tape video recorders, though more expensive, give them tough competition. Video disc systems that can record, however, are being developed for industrial use. Numerous laser experts predict that this may become one of the laser's most widespread industrial uses—to record information on video discs.

One thing is almost certain: Combined with other new technological innovations, such as TV stereo sound, an ever-increasing number of available channels, satellites, large screens and computers, lasers

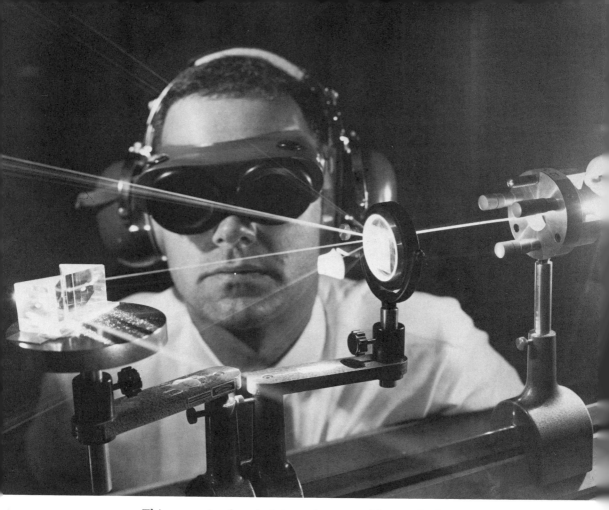

This argon-ion laser's intense beam could provide deep-space communication for missions to distant planets. Until we have such missions they'll still come in handy in computer processing and carrying voice and television signals. (Credit: Hughes Aircraft)

will participate in a video revolution in the coming decades. It is not difficult to believe this new age of video technology will have as much effect on coming generations as other communications changes had on earlier ones. Consider the differences brought about in the way we think, act, work and play due to the introduction of radio, movies, television and the long-playing record album—and that is only in this century. In other eras, the printing press, newspapers, magazines and readily available books wrought similar changes in the fabric of society and individual life. The technology of coming decades may be predictable; what changes of this sort may occur are not.

LASERS AND THE GLOBAL VILLAGE

Lasers digitally blipped through fiber-optic cables may one day create the communications miracles so often foretold by futurists, science fiction authors and pundits such as the late Marshall McLuhan. A prototype of one such system is already in operation in Higashi Ikoma, Japan.

Called HI-OVIS (Highly Interactive Optical Visual Information System), it is a computer-run optical communications center that links homes to the communications center.

Participants can shop by television at local stores, examine plane and train timetables, get the latest stock quotations, watch commercial broadcasts or check the weather report on their home screens. Users can take home-study courses complete with question-and-answer sessions and tests, in which a student gives answers via a hand-held keyboard.

In the future, according to the program's director, Dr. Masahiro Kawahata, the HI-OVIS system, which uses laser transmission of messages through fiber-optic cables, may give people immediate access to libraries, government officials, hospitals, out-of-town stores and various information networks.

Combined with computer response systems (such as QUBE, which is being tested in Ohio and elsewhere, satellite receivers that fit on the wrist, and numerous other advances), lasers will surely participate in the establishment of the global village.

Chapter 7

The Purest Light: Lasers in Industry

Laser crystal ball: Imagine, through the laser-lit ruby-red glow of the laser crystal ball, a rapier thrust of bright light boring a perfectly straight tunnel into the solid granite of a mountainside. Or, visualize the ultra-thin beam of an invisible CO_2 laser carving a line 10,000 times smaller than the distance halfway across a pinhead on a silicon chip that will prevent a car from running into a tree, another auto or a human being.

Think of a second. Count, one thousand, one. That's a second. Now, divide that by a trillion (10^{12} . . . ten followed by 12 zeroes, or 1,000,000,000,000). In the future, lasers firing pulses at that speed, already attained in the laboratory, will help engineers design microelectronic devices that make the transistor look like the Jolly Green Giant. "The future sizzles with expectations," says *Newsweek* magazine about lasers in industry.

Although more than 100 companies now supply various industries with lasers, and manufacturing, industrial and business uses of laser light increase every month, it was not always so. In 1978, *Newsweek* quoted one laser manufacturing official who ruefully noted that, "Industrial executives had seen too many Buck Rogers and James Bond movies. They were scared the lasers would vaporize their employees."

This was not a totally unwarranted fear. In one second, a powerful laser operating at one kilowatt can vaporize a steel ball bearing, or drill a hole through a worker's hand! Despite their science-fiction, Buck-Rogers image, lasers are by no means toys. Following several industrial accidents involving the use of lasers, the Federal Drug Administration, charged with regulating such equipment, issued standards in 1976 for protection of workers. Companies supplying lasers to industries also took pains to educate users in the benefits and dangers involved.

94

Currently, lasers are working to give manufacturers ultra-precise standards of measurement, unprecedented cutting and drilling power, light-speed quality control and are employed in a host of other ways. Doors may bear labels reading, "Danger—Invisible laser radiation—Avoid eye or skin exposure to direct or scattered radiation," but in spite of the risks, lasers are at work in plants and businesses throughout the world. They help shape ships, cars, jewelry and electronics components. They guide cutting, drilling and tunneling. They do things no other tool can do in the same time, offering the speed of light, a heat greater than that at the surface of the sun and a line straighter than any that can even be reasonably compared to ordinary rulers, tapes or surveying.

When lasers first appeared in the early 1960s, some pundits quipped that they were "a solution in search of a problem." A laser's properties vary a great deal depending on its size, construction, and lasing substance. Many uses required a good bit of engineering to develop the right combination of qualities for a laser to do a particular job. "It sometimes seems that each industrial application requires a different kind of laser," Dr. Arthur Schawlow has said. For this reason, lasers remain expensive. But in many industries, they do their job so well that the cost is recouped in efficiency and time saved.

Laser light has special qualities suited to industrial applications that no ordinary tools can match. Intense and highly directional, laser beams can be focused to needle a huge amount of power upon a pinpoint . . . or on an area much smaller. By focusing the laser's beam through lenses, just as you can tighten a sunbeam with a magnifying glass so that it will burn holes in paper, it is possible to make the coherent laser light concentrate its energy on an almost unbelievably tiny spot. Narrowed by optics, a laser can drill 300 separate holes in the head of a pin. You can get an idea of the concentration of energy in this tiny spot by considering what happens if you whack a piece of wood with a sledge hammer—at most a dent, but not much more. Hit the much smaller head of a nail with the same hammer, and you will drive the nail through the wood. Energy concentrated on a small point accomplishes much more than energy spread out over a wider area.

That's why coherent, tightly focused beams of laser light can do so much more than ordinary, spread out, incoherent light. You can get an idea of the difference in power between ordinary light and laser light by glancing at a household light bulb for an instant. The worst that will happen is that you will retain the image of the bulb on your retina for a few seconds after looking away. If you stared straight at a laser beam for even a fraction of a second, you could burn a hole clear

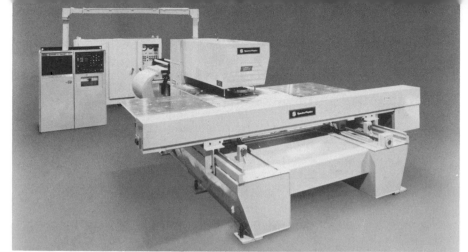

A

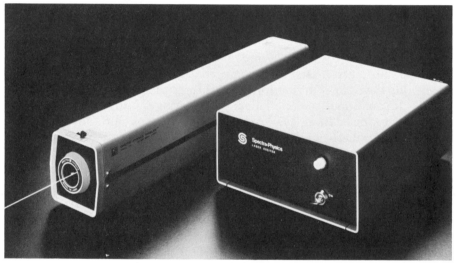

B

C

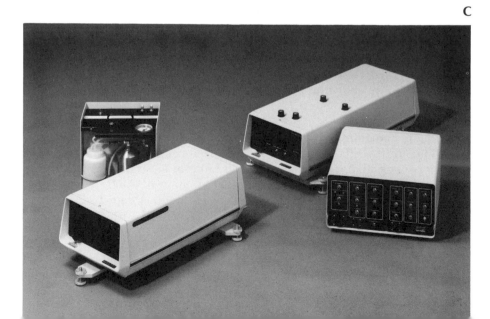

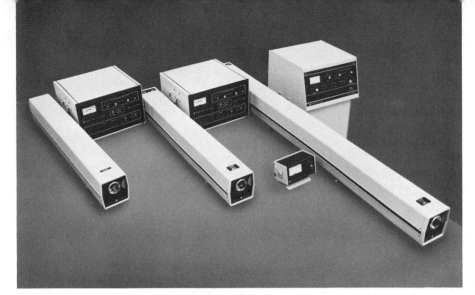

D

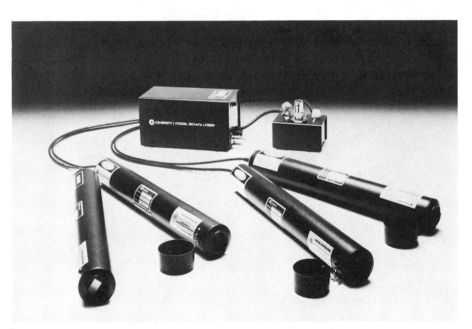

E

"It sometimes seems as if every industrial application requires a different kind of laser," states one scientist. The bewildering array available is suggested by the Spectra Physics and Coherent Radiation units shown here. (A) is a carbon dioxide metal-cutting laser controlled by computer; (B) a Spectra Physics Helium-Neon gas laser—note the needle thin beam; (C), a ring dye laser unit; (D)shows five- and 18-watt argon lasers; (E) shows components of a Coherent laser system. (Credit: Spectra Physics and Coherent Radiation)

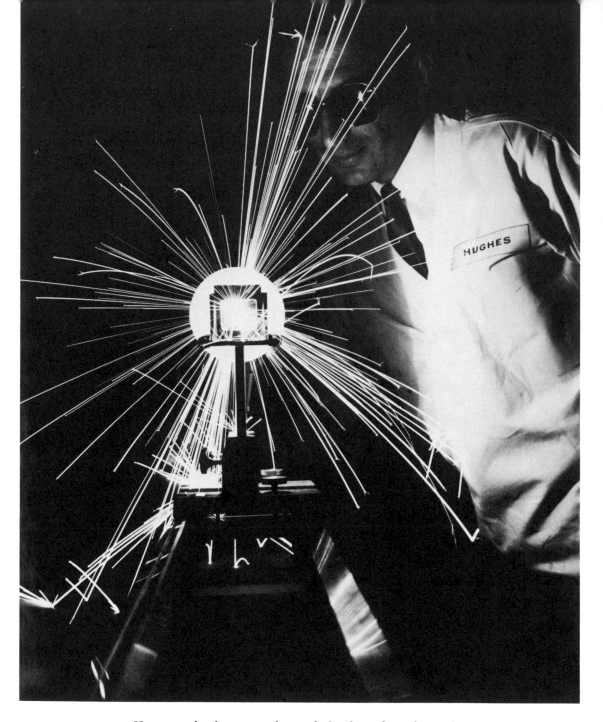

Here, a ruby laser punches a hole through a sheet of tantalum metal—an extremely hard material. The process takes less than a thousandth of a second. Laser cutting and drilling is gaining widespread industrial use because it is so clean, exact and fast. (Credit: Hughes Aircraft)

This Hughes Aircraft model X400 ruby laser is blasting a
hole through a stainless steel sheet at a temperature of 4,800
degrees F.

through your retina, resulting in permanent loss of eyesight.

DRILLING WITH LIGHT

Even a low-power, pulsed laser, which does not operate continuously, can zap holes in thin layers of anything—from diamonds, the world's hardest substance, to materials soft as rubber. Drilling holes in diamonds for wire inserts was one of the earliest uses of lasers in industry. Automatic lasers also drill ruby and sapphire watch bearings. They accomplish the job so quickly and efficiently that manufacturers found they could turn out more bearings than needed. Prior to the use of lasers, the same process was accomplished by a time-consuming procedure that required dipping threadlike wires in diamond dust to pierce the gemstones.

Lasers can drill soft materials as easily and quickly as hard, doing both better than other, conventional methods. They're used, for instance, to shoot holes in the nipples of baby bottles. But they poke precision holes in superhard metals used in aircraft, too. Lasers spike deep cooling holes in aircraft turbine blades to improve reliability and efficiency.

Lasers can spear quarter-inch armor plate in a fraction of a second (shades of *Goldfinger*) and lance thicker metal at the rate of a meter (slightly more than three feet) a minute. So many high school and college science students pierced razor blades with homemade lasers in science projects that one research scientist has suggested that a laser's power should be measured in "Gillettes instead of joules."

WELDING

Lasers offer numerous advantages in industrial welding applications. Welds hold things together (usually by joining two pieces of metal with hot lead or another substance). Lasers make welds stronger. Lasing is a non-contact process, so no pressure is applied to parts welded with laser light. Since laser beams are so compact, only a tiny amount of heat is transferred to the parts welded; this reduces the amount of metal distorted by heat, a problem that can weaken ordinary welds performed with torches. In ordinary welds, surface impurities can also weaken the bond. With lasers, this problem is eliminated. Pinpoint flashes of laser light can weld tiny bits of metal accurately and strongly.

Lasers are also used to weld two metals together or shave
tiny amounts of a material away in a process called micro-
machining. Here, you see an example of welding. (Credit:
Bell Labs)

Modern technology requires many tough, precision operations in which the laser plays a major role. Microminiaturized electronics components, which last years without difficulty, owe their reliability, and, in some cases, their existence, to the laser. Pinhead-sized transistors that need up to three exactly positioned welds are put into place by lasers that bond one-millimeter wires in a millionth of a second. Electronic memory chips that are small enough to be slipped through the eye of a needle can be welded by laser beams smaller than a hair's breadth. Because the laser can "select" its target, it can even weld metals through other materials without affecting them.

For instance, lasers have welded two wires four thousandths of an inch in diameter without affecting their insulation (the rubber around larger common wires is insulation). They can do the same thing through glass and other materials. Focused to a speck, the intense heat of lasers can melt almost any metal in a fraction of a second, including tin, zirconium, tungsten, columbium and many others. This quick, wide-ranging welding ability of the laser is especially useful in aerospace applications, where super-hard metal alloys are necessary, and welds must be as strong as possible. No one wants jet wings to fall off in flight, or satellites to crack in space.

Although laser systems cost up to $500,000, the gain in quality, accuracy and higher production rates often make the investment worthwhile. In some cases, laser welding is the only practical method available to do certain jobs. The laser's ability to weld completely different metals is particularly helpful. For example, electronic and higher production rates often make the investment worthwhile. For example, electronic controls in cars require the use of nickel-plated or stainless steel wire rather than conventional types, and only lasers act on these materials effectively without affecting delicate sensors and other equipment. In microelectronics, old methods of arc-welding cracked 50 percent of the components—laser welds, because of their speed and pinpoint accuracy, do not make cracks. This results in an enormous savings both at the manufacturing end and for consumers, who receive more reliable products and thus ride safer vehicles, be it cars or jet planes.

CUTTING AND HEAT TREATING

While lasers shown in popular films and other media are frequently flashy units that slice through thick armor plate, their accuracy and speed is equally useful in cutting softer materials. They cut clothing,

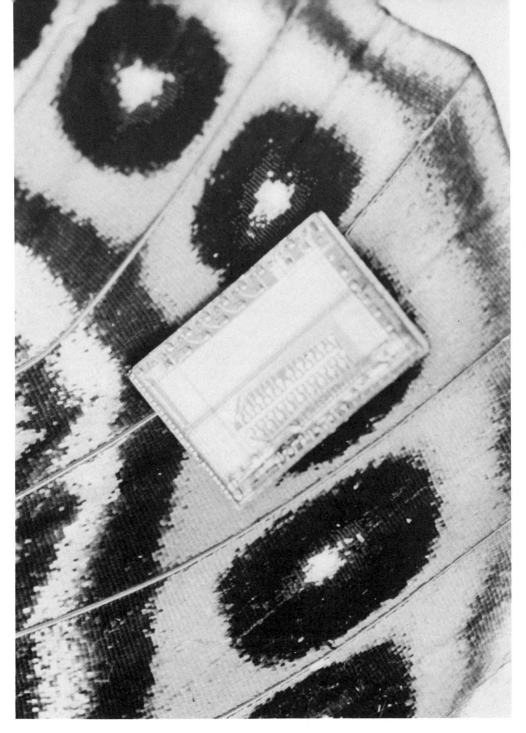

IBM's densest memory chip can store 72,000 bits of information and is shown on a butterfly's wing. Without lasers, reliable electronics of this size would be impossible. (Credit: IBM)

rubber, plastic and wood as well as metals.

Powerful CO_2 gas lasers etch narrow slots, fine lines and intricate patterns better than any conventional tools. Since plastic and rubber absorb the CO_2's infrared light easily, these lasers are especially suited for fabrication of a variety of rubber and plastic products. Some units slice plastic at the rate of a foot per second. Thin plastic decals for cars are one product often cut by laser beam. Laser engraving of wood plaques, trophies and novelty items is becoming standard procedure. Nylon belts sliced and sealed by lasers don't have ragged edges to ruin clothes.

THIRD WAVE TOOL

Future Shock author Alvin Toffler believes that lasers, along with computers and the electronics revolution, are one of the major tools helping to create what he calls the *third wave*. In his book of the same name, Toffler asserts that lasers are playing a role in shaping a brighter, happier future in which man will have more control over his life. As an example, he cites the use of the laser gun used to trim clothing.

"The new laser machine operates on a radically different principle" from either the *first wave* method of hand-making clothes, or the *second wave* process of industrial mass production, Toffler wrote. "It does not cut 10, or 50, or 100 or even 500 shirts at a time. It cuts one at a time. But it actually cuts faster and cheaper than the mass-production methods employed until now." By reducing waste and the need for large inventories because garments can be made quickly, laser machines make it possible to fill an order for even a single garment economically. This suggests, states Toffler, that mass-produced, standard-sized clothing may one day disappear.

Future shoppers may tell a manufacturer their measurements by phone, or give a computer a visual reading through video scanning. Then, the computer will guide the laser to custom cut the desired garment. This, notes Toffler, is a return to the earlier method of custom making all clothing, but on a high technology basis.

CARS, SHIPS, PLANES

America's largest automobile manufacturer, General Motors, has been described as "the biggest fan of lasers in Detroit." GM hardens the metal housings of power steering mechanisms with seventeen lasers that cost two million dollars. When combined with computer controlled systems, lasers can operate completely automatically.

The laser heat-treats steel shafts and other parts better than conventional methods for several reasons. Major distortion problems are avoided because the laser heats only the selected target, such as a protective surface, and not the base material. Lasers turn themselves off when the job is done, saving energy. They can carve helical patterns in hardened surfaces that make them wear longer by trapping grit and other foreign substances in the helical grooves, thus preventing damage to the hardened surfaces. One of their newest applications is in bonding alloys to metal surfaces, increasing their resistance to wear, heat and corrosion. For this reason, lasers frequently heat-treat various parts of cars, ships and planes that must undergo high stress-and-wear conditions.

The process of hardening materials with heat is called *annealing*. A spokesman for one of the major optical industrial magazines dealing with lasers says that "Laser annealing is one area in which scientists and engineers are coming up with something new just about every day. It will be one of the main industrial uses of lasers in the future."

MEASUREMENT AND DIAGNOSTICS

The intense, monochromatic, ultra-narrow lines of the laser beam form ideal measuring tools. Light rulers guide bulldozers, cutting machines in factories, pipe laying, tunnel digging, surveying and alignment of machine tools. Visible helium-neon gas lasers produce a beam of extraordinary purity that is highly directional. It can measure changes in movement as small as ten millionths of an inch. If a 12-inch ruler were the Empire State Building, ten millionths of an inch would be the thickness of a piece of paper. Keeping his work aligned with a laser beam, a machinist can tell if the object he is shaping changes position by less than a hair's width. Unlike such alignment references as string or wire, laser beams do not sag. A trench, tunnel or pipe aligned by laser has an accuracy that varies less than one millimeter in 100 feet.

One laser manufacturer, Spectra Physics, boasts that its products "can align underground pipes and sewers, survey construction sites, level floors and ceilings, and plumb walls faster, cheaper and more accurately than the conventional engineering tools used since the days of the Pharaohs." The company is also proud of its Laserplane

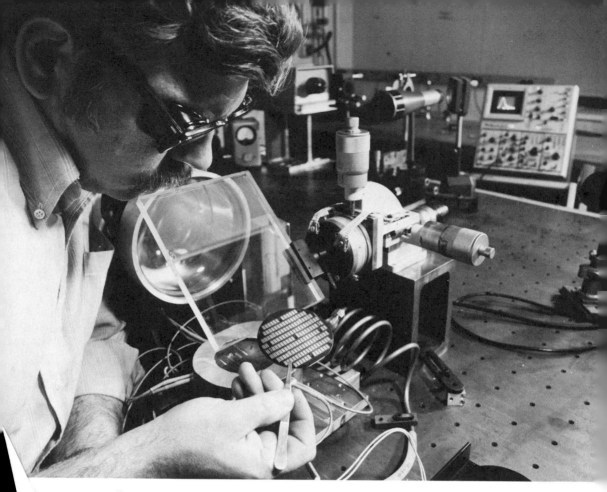

One laser scientists describes laser annealing as the jazziest field in industrial laser applications. Annealing is a heat treatment that can reduce even further the size of miniscule semiconductor devices. The photo shows a silicon wafer ready for laser annealing. The angled glass to the left of the wafer directs the laser light down onto the wafer. (Credit: Bell Labs)

(trademark) system. "These guidance and control systems for earth-moving equipment permit relatively unskilled operators to level and contour farmland, lay runways and highways in less than half the time required by conventional methods," claims the corporation's annual report. Its Laserplane system is used extensively in major grain producing countries and developing nations. There it enables technologically unsophisticated workers to prepare large land areas for flood irrigation of crops.

This photo shows the effect of laser annealing on the tiny widths (called domains) in a magnetic bubble memory. The center domains have been treated and are reduced from 16 to 7 micrometers. That means that future electronic devices like calculators, computers or even spacecraft can grow even smaller. (Credit: IBM)

Construction equipment guided by laser beams is used to level building sites, drill tunnels, and lay pipe. The unwavering beam of light used to guide this bulldozer's blade assures unprecedented accuracy. (Credit: Spectra Physics)

Another laser company, Coherent, Inc., makes a "Flatness Analyser" that measures silicon wafers and masks for the semiconductor industry. Printing increasingly complex semiconductor circuits on ever smaller silicon wafers requires them to be flat within a ten-thousandth of an inch.

In the optics industry, many high-resolution lenses must be measured to an accuracy of a millionth of an inch, and here too the laser does the job better. Ford Motor Company scans the steel shafts of power steering units with lasers to check their smoothness. A Uniroyal factory checks specifications on tire molds with lasers.

At Sandia Laboratories in Livermore, California, researchers probe the way internal combustion engines work with lasers, hoping to design engines that require less fuel. Surprisingly, many facts about how gasoline-air mixtures behave inside of engines are still incompletely understood. The scientists at Sandia explore the chemical changes going on inside pistons by making laser shadowgraph movies, speed measurements and chemical analyses. Unlike ordinary engine diagnostics, the laser beams do not change what's going on. Examining the characteristics of the swirling gases inside the engine via laser beam, making laser movies of combustion flames and measuring the speed of various processes just might help major motor companies build a better product. In the not-too-distant future, the car you drive or ride in may not only have parts made by laser, but also an engine designed by laser.

RETAILING, PRINTING, PHOTOGRAPHY

Laser label readers are quickly becoming one of the most common uses of the light fantastic. Many grocery stores have installed the quick, automatic laser scanners that rack up an item's price when it is drawn across the beam. They haven't caught on everywhere, though. Consumers raised a bit of a storm about them when they were introduced on the West Coast because they suspected stores could raise prices more quickly and less obviously with them.

One IBM system can read a Universal Product Code anywhere within 180 degrees of its holographic, wrap-around beam. All checkers need do is tip the product to the proper angle and the laser beam automatically records the price. Large store chains like this automatic equipment for several reasons. It increases the labor pool on which they can draw. Inexperienced checkers quickly become as fast as those who have been on the job for years. This substantially reduces customer waiting time at check-out lines. Also, supermarkets operate on a low profit margin—as little as one percent of gross sales—and underestimation of prices by checkers using regular cash registers can cut that profit in half. Laser scanning systems reduce or eliminate the problem.

In 1979 alone, Spectra Physics, which developed laser supermarket systems in the mid-1970s, installed more than 7,500 units. The company predicted it would sell 20,000 more by 1981. Still, consumer and labor resistance to the devices continues. Fewer than 10 percent of major supermarket chains will have the systems by the end of 1981 at current installation rates.

One of the fastest growing areas in which lasers have industrial applications is in printing. Color photographs reproduce faithfully when scanned by argon ion lasers. Laser engraving systems work by vaporizing bits of a rubber roller to match the patterns of light and dark in artwork, leaving a raised image that's used to transfer the art-

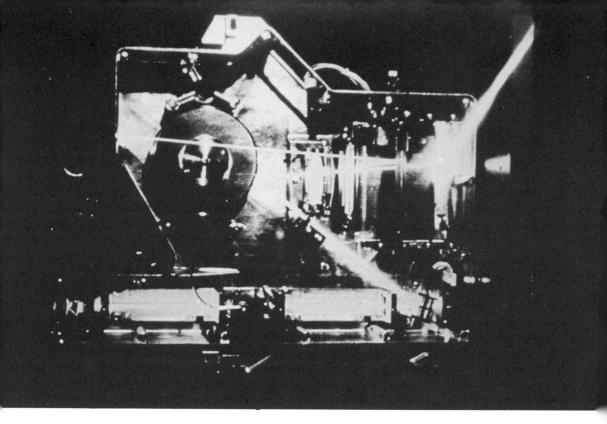

The inner workings of the IBM laser printer. The laser light
is deep red. (Credit: IBM)

work's image to paper. And, combined with computers, lasers etch
printing plates more quickly and precisely than other methods, providing better looking pictures and copy.

IBM has even introduced an office machine that not only prints
with a laser, but also receives and transmits documents electronically
over ordinary telephone lines. It will also link word processing and
data processing units. "We believe this product represents a signifi-
cant evolutionary step toward the much discussed office of the future," asserts IBM Vice President J. Richard Young.

The IBM laser printer spits out up to 13,360 lines a minute (or an
incredible 45,000 characters a second). It permits use of four different
type faces, which can be mixed in any sequence to separate various
kinds of material on a page. Its low-power laser "prints" character
images on a photoconductive surface covering a rotating drum. As the
drum turns, dry, ink-like powder that adheres only to these images is
transferred to paper, supplying the printed text. Forms with complex
designs can be projected onto the printing drum by flashing a strobe
light through a film negative. The laser itself forms less complicated
designs.

One thing is certain. The laser, once described as a "solution without a problem," is solving problems galore in industry, business and retailing. In combination with the other tools of modern electronics technology, lasers are helping to create the future. And, if these peaceful, constructive applications of lasers are sometimes overshadowed by its reputation as a "death ray," it may be because we are only at the threshold of a better world. The pure bright beams of lasers may help light the way as we step into tomorrow . . . and the day after.

Laser Crystal Ball

In the near future, you may be able to practice your golf swing, softball skills or tennis serve via computer-controlled laser machines. Combining a holographic video disc playback with laser scanning and sensing devices, it will be possible to perfect your skills in the comfort of your living room.

Essentially, such systems would work by recording your performance through a video camera. It would relay its information to a computer-operated photosensing mechanism. The imaginary ball you strike, whether a golf ball, tennis ball or Ping-Pong ball, will pass through photocells into a net. A laser scanner would "read" the video-tape and locate your ball on a golf course, tennis court or softball field.

While this laser application is still in the future, others that sound like science fiction are already in use. Crime lab detectives beam argon laser light at objects handled in thefts or murders to reveal fingerprints left years before. By focusing on oil residue left by fingers, the laser can outline fingerprints even on objects that have been scoured.

Jewelers streak laser light through diamonds . . . or fakes to identify the gems from the glass. True gems break the laser light into a unique, identifying pattern, thus making identification of a jewel simple.

SAVING THE OLD WITH NEW TECHNOLOGY

When Dr. John Asmus, a San Diego research physicist, went to Venice in 1971 to make holographic records of that city's famous statuary, he stumbled upon a discovery. Not only could lasers preserve the rapidly

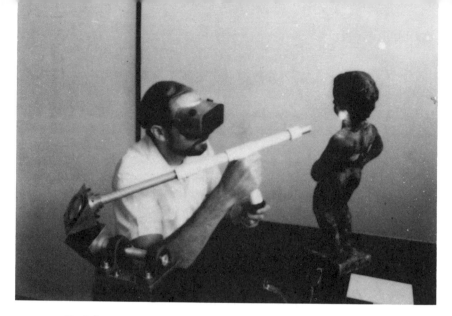

Dr. John Asmus uses laser light to clean a valuable work of art. (Credit: John Asmus)

deteriorating works as holograms, they could also clean and restore the sculptures themselves.

The black crust on the Venetian statuary, caused by years of air pollution, salt-water breezes, flooding and pigeons with no respect for art, burned away under the laser's hot beam. The white marble underneath did not absorb the laser's energy and remained unharmed.

Since this discovery, Asmus has become an expert in cleaning and restoring everything from paintings to buildings with lasers. "A variety of things come to me from art restorers," Asmus relates. "About the only thing they have in common is that everything else tried has failed. They turn to my techniques as a last resort. Generally, the work has to do with selectively removing a disfigurement, which could be graffiti, corrosion, overpaint or oxidation. The objects range from books to bronze statuary to stone monuments. Usually the art work itself is fragile and barely holding itself together. The material you'd like to remove is more tenacious, harder or chemically resistant."

Lasers, however, strip graffiti from New York subway walls, peel corrosion from bronze and stone, or cook the mildew from rare books with ease. Depending on the job, laser light can actually change the chemical bond holding unwanted material to a surface, vaporize dirt or chip it away.

"Cleaning things accounts for about three percent of the gross national product," Asmus notes, "from cars, ships and bridges, to buildings." To get the job done, Asmus has used all sorts of lasers, "ruby, YAG, argon, CO_2, excimers and even ordinary or non-laser flashlamps."

Some of the work he's done is impressive. After vandals defaced irreplaceable Indian rock paintings in Utah in 1979, Asmus was called in to try laser restoration. With brushes and kitchen cleansers, the vandals had tried to erase the Indian pictographs. The burning desert sun baked the cleanser onto the sandstone wall, hazing the previously brilliant drawings.

Asmus pulsed laser beams at the cleanser's residue, returning some areas of the pictographs to full brilliance. In addition, other paintings were discovered under the originals that are completely intact. "We didn't even know they were there," said a park ranger.

In other jobs, Asmus removed nine layers of paint from large historical murals at the California State Capitol rotunda. He removed hog fat from cracks in the marble structure of a 600-year-old Italian cathedral. Well-meaning workers applied the lard during the previous two centuries in attempts to preserve the marble, but it actually hastened decay. In Florence, Italy, Asmus zapped away tempera overpaints on 14th-century frescoes. He has also cleaned Bavarian wood carvings and medieval leather-bound books.

His work with lasers is not limited to directing the laser division of Maxwell Labs in California or cleaning and restoring operations. Asmus also consults with the Defense Department on laser weapons technology, and has used them in work for the space program.

He first stumbled upon lasers while working on his doctoral thesis in 1960. "I'm a plasma physicist," he explains, "and I was trying to do a plasma measurement, but was having trouble getting enough light intensity from normal incoherent sources. A friend of mine told me about the first working laser at Hughes, and I realized that would ease my work dramatically. So, I promptly built one of the first lasers then in existence."

Later, he began doing work for the space program, using a gigawatt laser to hurl micro-meteorites at high speeds and check the craters they made when they struck targets in the lab. "In the early days of the space program there was a lot of concern about these micro-meteorites and the damage they might do to spacecraft. I think I held the world's land speed record for about 10 years doing this," he quips.

In the late 1960s, Asmus began research in extracting oil from shale with laser radiation. He was also involved in projects to gasify coal and chemically synthethize alcohol and other substances with lasers. By 1970 he was in Washington, D.C., coordinating Defense

How It Works: Lasers in Industry

Most cutting, drilling and welding with lasers rely on the continuous high-power output of carbon dioxide (CO_2) units.

Drilling   Lasers actually burn rather than drill holes in substances, from the toughest steel alloys to baby-bottle rubber nipples. The laser concentrates its energy in a tiny beam that can be further narrowed with lenses and other optics. When it strikes the surface to be drilled, the laser's beam raises temperatures at a small spot (as little as hundredths of an inch, though larger holes are also drilled if necessary). The concentrated laser energy burns at the rate of 10^{10} (10,000,000,000) degrees-per-second temperatures. This heat vaporizes the surface to a depth of only a few microns. Extremely high pressure builds in the hole, shooting the material out of the depression in a plume. Often, a gas jet will be used to speed the process by blowing away vaporized debris and particles.

Cutting   The same basic procedure is used to cut, trim, engrave or scribe materials ranging from steel to plastics. One of its advantages is that when cutting synthetic materials such as plastics and laminates, it seals edges automatically, producing very smooth surfaces and saving manufacturing time. Since laser cutting is a non-contact process—the light beam applies no pressure to the surface being sliced—it parts compressible materials (such as foam rubber) much more effectively than hot-knife methods. Lasers are also particularly effective tools for machining hard, brittle materials such as ceramics and glass. Hardness does not affect the laser's cutting ability. Normal diamond drills, special saws and other machines used to work on these materials frequently break or must have blade or stylus replacements. Lasers thus reduce down-time and eliminate the need for costly industrial diamonds and tool-steel blades.

One of the more unusual applications of the laser's cutting talents is in the record business. A & M records has intricate designs laser-etched into the vinyl of its discs (LP's by the groups STYX and Split Enz have them). These unusal designs are not only visually interesting, they also serve as "watermarks" that help prevent record piracy.

Measurement   It is virtually impossible to match the speed, straightness and effectiveness of laser light as a measuring tool. The wavelength of visible light is about the thickness of a soap bubble, so it can register tiny changes in position, or align pipes, tunnels and machines with unprecedented accuracy.

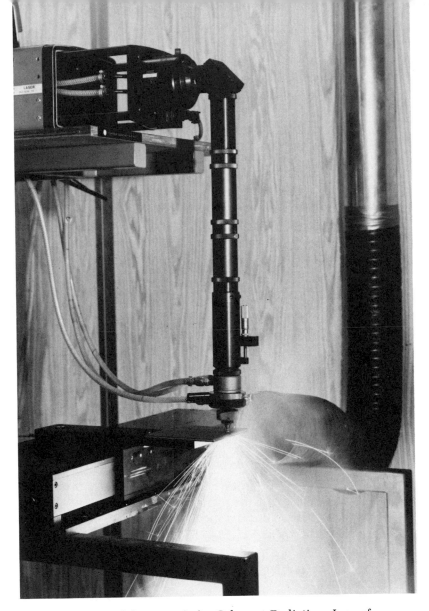

An industrial laser made by Coherent Radiation, Inc., of California, is shown here boring through a metal plate. Note how close to the target the laser is placed. Cutting and drilling lasers can raise target surface temperatures 10,000,000,000 degrees Centigrade a second—hotter than the temperature at the surface of the sun. This blasts out a tiny crater only a few microns deep, but huge pressures build underneath the beam and explode additional material outward. Gas jets are often used to help wash this material away. You will observe that most photographs of lasers blasting metal and other materials show the explosive plume of incandescent debris spraying outward. (Credit: Coherent Radiation)

Laser scribing (etching cut patterns in ceramics, metal, electronics components, wood) is widely applied in industry. The rooster-tail pattern shown here is excess material vaporized by the light stylus as it does its work on Bell Telephone microprocessors. (Credit: Bell Labs)

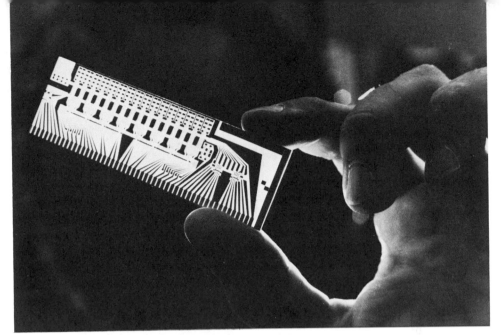

This microcircuit pattern was cut directly onto this circuit board with a laser. (Credit: Bell Labs)

Department laser weapons programs. While there he met an oceanographer who told him about how Venetian artwork was falling apart and not much was being done to save it, leading to his holography project.

Today he spends about 10 percent of his time on art restoration. "About 50 percent of my time has to do with Defense Department matters, the remaining 40 percent in industrial applications of lasers. The most popular use in industry now is laser annealing of semiconductors. That's a very jazzy field." Using lasers to heat-treat these tiny electronic devices is expected to lead to yet further miniaturization of the multitude of modern tools, toys, communications devices and computers which rely on them.

Where will lasers be used most in the future? "The one use that probably overshadows all the others combined is laser-induced chemistry," Asmus believes. This involves using laser light to make certain chemical reactions happen. "All kinds of people are doing this in the lab, but not in industry. I think that in a decade or t wo we'll see enormous use of this throughout the chemical industry."

Asmus confesses that, like many modern scientists and inventors, he was at the right place at the right time. "I look upon lasers as that thing which has made my life exciting. If I had mapped out what I thought would be an exciting life for myself back in 1957-8, I think it would pale in comparison to what really happened through chance and serendipity in the last twenty years. The fact that the laser came along at that time is largely responsible for the surprises and exciting things I've experienced."

Chapter 8

To Light the Darkness: Lasers In Scientific Research

Firing pulses of laser light so quickly that a speeding bullet or orbiting spaceship would appear motionless in the time it takes the laser to flash a million times, scientists are probing the world of the atom. On the other side of the superfast, ultra-slow time scale that nature sometimes uses, lasers measure speeds as slow as one seventh of an inch per hour to provide researchers with new information about barely perceptible processes in human cells.

By observing scattered light from lasers aimed at various substances, still other scientists are developing extremely sensitive methods of detecting what is in our air, food and water. With the aid of laser light, they see as little as a single atom, revealed by the way it spreads laser photons that strike it.

Whether drilling through steel plates in a shower of sparks, spearing missiles in flight or slicing a tumor from a man's brain, a laser is visually impressive and exciting. But scientists throughout the world are using lasers for an equal, if not more important, purpose: probing the secrets of nature.

"Laser technology harnesses light in a more sophisticated way than it could be harnessed before," says Dr. John Asmus, who has been involved in nearly every aspect of the field. "The invention of that first gadget we call a laser led to a great many innovations, changes and expansions in the way we use light. It's comparable to the harnessing of electrical energy or nuclear energy rather than to, say, the transistor or the computer. Laser technology represents a field concerned with control of a fundamental form of energy."

This "control of a fundamental form of energy" gives scientists a research tool that will undoubtedly make many "gadgets" possible, just as control of electricity did. By lighting the worlds of the supersmall, ultra-fast and oh-so-slow lasers help scientists seek basic knowledge that accumulates in tiny bits until a host of applications

118

A long-life argon gas laser developed by RCA scientists for NASA. Here they are measuring the diffused beam. NASA developed the device for use in space tracking and communications. (Credit: NASA)

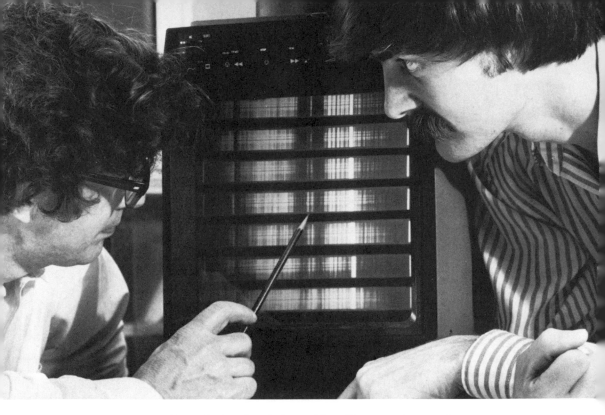

Dr. Peter P. Sorokin (who, with colleagues, invented an early four-level laser), left, and Dr. Donald S. Bethune study a series of spectra showing the progress of a chemical reaction. A new laser technique developed by Drs. Sorokin, Bethune and their colleagues at the IBM Research Center in Yorktown Heights, N.Y., permits these "spectral snapsnots" to be made of events that last only a few nanoseconds. (Credit: IBM)

appear. Sometimes the results in practical terms, such as a better mousetrap, do not appear for decades. Others seem to pop into modern life overnight, as did the silicon chips that made calculators and electronic games possible.

LIGHT MATTERS

The invention of the spectroscope led to a remarkable discovery that greatly enlarged man's understanding of light and matter. Indispensable in astronomy, physics and chemistry and many other branches of science, the spectroscope reads the "fingerprints" of nature's building blocks, the atoms.

A spectrum is the multicolored band of light you see in a rainbow. Ordinary light, such as that from the sun, a household bulb or other common source, will spread into the entire rainbow band when passed through a prism, from violet to red. When a given substance is heated to produce a gas or vapor, it will radiate and absorb light only over a narrow band of frequencies. By shining a light through the vapor, then projecting that light through a prism onto a viewscreen, a characteristic spectrum will be seen. Since the substance absorbs certain frequencies of light, those frequencies will appear as dark gaps on the viewscreen. The pattern of these dark lines on the spectrum of colors, made as the unabsorbed light is split into its remaining parts by the prism, is always the same for a given element. This pattern of lines reveals each element's spectrum of frequencies; for each element, the pattern is as precise and individual as a person's genetic code. Scientists have measured the spectrum of every element. They have tables showing where each line appears for a given element, even when the lines are no more than ten billionths of a millimeter apart.

Spectroscopes, like lasers, work because each element has a unique set of energy levels. Each atom is like a house with a certain number of floors, but no stairway. Electrons can travel from one floor to another on an energy elevator that stops only at the floors, never in between. As the electrons go up they absorb energy only at frequencies associated with each floor. When they descend to lower floors, they return energy, radiating it at the same frequency. Since each atom has a specific number of floors, a spectroscope "reads its blueprint" by revealing what frequencies of light it absorbs.

LASER LEAP

"Lasers have become indispensable tools for modern spectrography," notes Dr. Arthur Schawlow of Stanford University. "The fact that laser light is monochromatic (a single color) has allowed us to investigate much more deeply the nature and properties of matter and its interaction with light." With lasers, it is possible to determine the spectrum of a substance by varying the frequency of the beam and observing which frequencies are absorbed. This eliminates the need to analyze the colored spectrum produced by a prism, simplifying the procedure and increasing accuracy.

In an article for *Science* magazine, Schawlow noted that the special "purity" of laser light "makes possible unprecedented resolution of fine detail." It eliminates undesired broadening of spectral lines, which can be confusing. "With lasers, we are finding possibilities and many new ways to probe deeply the nature of matter."

The consequences of work in laser spectrography are enormous and wide-ranging. "We've found we can observe, by the scattered light from a laser, as little as a single atom," Schawlow explains. "So, we are developing extremely sensitive methods of detecting a small amount of something. We can certainly see that not far ahead we will be able to detect one single cancer-causing molecule in a quart of milk. If you do that, you won't be able to talk about anything being really pure. We may find there is no such thing. We may have to decide what degrees of contamination are tolerable.

"On the other hand, we might be able to detect traces of substances that might be helpful, things we never suspected were there, but are essential to life."

On the way to being able to "measure everything," scientists are also using the laser to simplify viewing spectra, revealing in the process new information about atomic transition states (energy level changes). "In 50 years of work," says Schawlow, "scientists were only able to identify six electronic states. We've found 22 more. It gets a little technical, but essentially, we can use the selective waves of the laser to label a particular level (transition state) and measure only the ones that are there."

At Colorado University, scientists are putting laser spectrography to work searching for "exotic" atoms that may have been "left over" in small numbers by the big bang that astronomers believe created the universe. Lasers are also involved in the tests for neutrinos, the elusive subatomic particles. In both of these instances, laser technology may produce answers to basic scientific questions about how the universe began, what it is made of and how it may die. For centuries, man has sought the answers to these questions through philosophy, religion and science. It is fitting—if also somewhat ironic—that the answers, long shrouded in darkness, may, in part, quite literally come from light.

STUDYING THE SUPERFAST

On the tiny level of the atom, things happen in billionths and trillionths of a second. Scientists call a billionth of a second a *nanosecond*; a trillionth is a *picosecond*. To study what happens at subatomic speeds, researchers use pulses of laser light so rapid that they are measured in picoseconds.

Talking about billions and trillions (or billionths and trillionths) of something is much easier than actually visualizing those numbers as units. Just how short a time is a picosecond? To provide an idea, consider something usually thought of as moving very fast in the macroscopic (large) world we normally experience: a space capsule traveling five miles a second (18,000 miles an hour or 72 percent of the way around the earth in one hour). In a millionth of a second, the space capsule would travel less than a third of an inch. In a nanosecond, it would not move three ten-thousandths of an inch. In a picosecond, it would progress less than three ten-millionths of an inch.

Or, think of it this way. Light travels at 186,200 miles in one second (nearly 300,000 kilometers). Although light moves in a straight line, if it could round corners it would circle the earth seven times—and part of an eighth—in one second. In a nanosecond, it travels a foot. In a picosecond, light, which moves quickly enough to seem instantaneous in the large-object macroscopic world, travels a mere 1/80th of an inch.

Firing ultra-short laser pulses lasting only picoseconds, scientists at Bell Labs, IBM and the Naval Research Lab among others, are learning new details about what happens within the atom's tiny time scale. At Bell Labs, Dr. C.V. Schenk heads a team pushing into the borders of the unknown with these superfast flashes from organic dye lasers. Tunable overa a broad band of light frequencies, these lasers can generate very short pulses indeed, on the order of .09 picosecond (or .000 000 000 000 000 9 of one second). Splitting the pulse in two, scientists actually use it to "measure itself," explains Schenk, "when we bring the pulses back together."

In one way or another, time always involves distance or space, as well. Think of how we measure it in the world we are familiar with : Watch hands that travel the distance between numbers on the face; the pendulum that controls the hands by moving back and forth over a certain distance at a precise rate; the earth turning on its axis or orbiting the sun. By putting distance (very tiny, certainly) between the laser picosecond pulses, scientists can measure movements much faster than ever before possible.

Already, Schenk's team has discovered that in one semiconductor substance, gallium arsenide, subatomic particles behave differently than previously thought. Their speed, over very short distances, is not slowed by collisions with solids. "These are measurements designers will have to consider when making very small, very

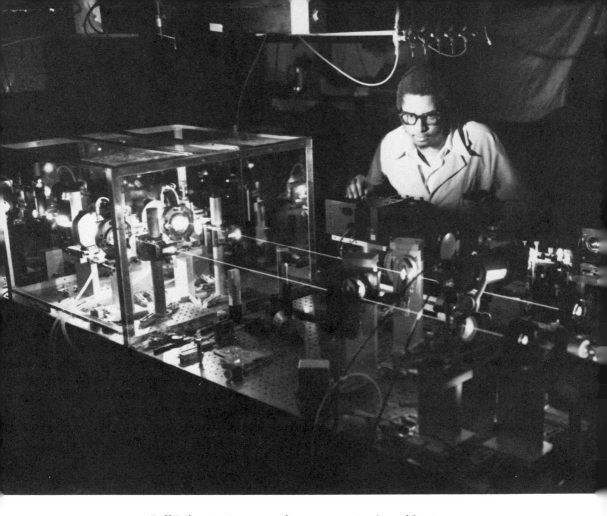

Bell Laboratories researcher uses a pair of tunable picosecond dye lasers to study the ultrafast electronic response of semiconductor surfaces. These and even faster spikes of laser light help scientists explore the limit and frontiers of knowledge, and design new electronic devices that will probably approach the practical limits of physics. (Credit: Bell Labs)

high-speed semiconductor devices." As the size of these electronic components shrink, an understanding of exactly what happens to make them work becomes increasingly important. The next generation of such devices will be less than a micron in size (one millionth of a meter). "Our discoveries may cause important changes in the way small electronic devices are designed," Schenk feels. "The observation that electrons in microcircuits first speed up, then slow down, opens an avenue for the development of new devices."

Another use of the ultra-short pulses is in studying fundamental chemical processes and very fast reactions. Some of these chemical reactions, such as combustion in a gasoline engine, occur so quickly they are poorly or incompletely understood. Rapid flashes of laser light help scientists "read" what is happening chemically.

"On the order of picoseconds," Schenk explains, "molecules change their configuration (shape), they interact with each other, collide in solvents, exchange energy and do lots of things. The very short laser pulse is an important probe because it allows us to make measurements on this time scale."

By flashing superfast laser pulses of different colors to start and stop electrical signals, Bell researcher David Auston created electrical switches from 10 to 100 times faster than transistors. Auston asserts that devices as fast as one picosecond (remember, it takes 80 picoseconds for light to travel one inch) are possible. A switch of this kind would dramatically increase the speed and reduce the size of computers, communications equipment and other semiconductor devices.

Auston's switch works by focusing beams of laser light on a piece of light-sensitive silicon (silicon is one kind of semiconductor). Pulses of two different colors, one to turn the switch on, the other to turn it off, switch electrical signals through the silicon crystal in picoseconds. Because no conventional instrument could measure switching times that fast, Auston invented a special technique to do so. He connected two switches in tandem and used the second to measure the speed of the signal in the first.

THE FASTEST CAMERA

Stop-action photography is accomplished by taking many still photographs of an event at high speed. The faster the shutter clicks, the more different movements it will record on film. Ordinary high-speed photography, though it will capture the fastest race car moving at top speed without a hint of movement, is not fast enough to analyze such events as a bullet penetrating body armor. At the U.S. Naval Research Lab inWashington, scientists built a camera system with lasers that can.

The system's guts are six ruby lasers pulsed in 30-nanosecond flashes through optical fibers. Because it stops action at these 30-billionth-of-a-second intervals, the system reduces blurring 20 times better than any other method. Sharper pictures of very fast events are

Dr. Renne S. Julian, senior scientist at Hughes Aircraft Company, sights through alignment device of a laser range-finder that beamed a high-power laser pulse to the moon through a telescope. (Credit: Hughes Aircraft Company)

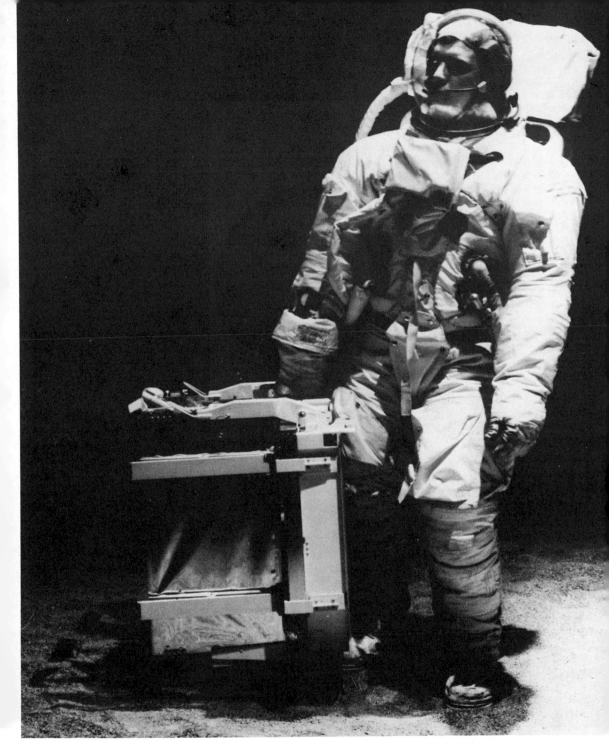

This is the "Laser Ranging Retro-Reflector" astronauts placed on the surface of the moon to be used as a target for earth-based laser systems. (Credit: NASA)

thus possible. Don't look for this device at the local camera shop, however—at least not for quite a while. It cost $250,000.

Biology is yet another field in which the speeding spikes of laser light measure what nothing else can. In one experiment, Bell scientists studied the way light cures babies born with an enzyme lack. Some babies, born without enzymes that break down impurities in the blood, are jaundiced (yellowish in color). Shining light on them can cure the ailment. "It's an example of a light process, and we looked at the fundamental physics of how it happens," Schenk reported.

GO SLOW

In medicine and biology, measuring extremely slow movements can be as important as measuring those extremely fast. With the technique called *laser doppler velocimetry (LDV)*, researchers can observe the slow poke motion of protoplasm in cells and other body functions. LDV works on the same principle as radar. Light reflected by a moving object scatters in a different frequency. The amount of frequency shift is directly proportional to the speed of the moving object. This "doppler shift" occurs with sound waves as well as light waves. You can hear it in the change in pitch of a police siren or car horn of a vehicle moving toward or away from you, and even in the scream of someone falling a long distance.

SPOTTING PROBLEMS

By focusing the laser's beam to a microscopic point, scientists can measure tiny changes when its scattered light is analyzed. The technique is being used to identify male infertility problems (difficulties in having children). Finding out how much sperm is there and how slowly it moves can aid in selecting a treatment. In hospital labs, the technique has speeded up a test that identifies cancer cell types. The test once took hours, and now requires only seconds. This can help doctors decide on proper treatment more quickly, which is extremely important in treating cancer.

One of the major benefits of this technique is its ability to measure blood flow rates. It produced the first accurate picture of blood flow rates in the human eye. Blood flow rates can reveal: how well blood circulates in vital organs affected by hardening of the arteries; how deep burned tissue goes; how well skin grafts are doing; and other important information.

UVASERS AND RAYSERS

Efficient lasers that operate at shorter wavelengths—ultraviolet and soft x-ray frequencies—are a possibility scientists see in the future. C.K. Patel, inventor of the carbon dioxide laser, notes that "One can visualize a number of applications for them. Three-dimensional holographic pictures of viruses might be possible with x-ray lasers." Tiny x-ray waves would be necessary to outline a 3D image of a virus. A typical virus is only 100 angstroms long. An atom is approximately one angstrom; you can see how tiny this is when you realize that the grooves on a long-playing record are separated by about 100,000 angstroms.

TRIGGERING CHEMICAL REACTIONS

Many laser specialists, from pioneering scientists who have worked with the devices since 1960 to editors of magazines specializing in laser technology, feel that "laser-induced chemistry" will be one of the major future applications of coherent light. Dr. Asmus (*see* Industrial Lasers, page 111) cites this field as an example of where lasers will be put to work more often in the future. "Laser-induced chemistry is one field currently in an embryonic stage. People are doing it in university research labs, but not in industry. I think it will be very important in coming years," Asmus says.

Both in research and industrial chemistry aimed at producing consumer products, scientists need two things for successful work: They must understand the chemical structure of the substances used, and they must be able to create chemical changes at a high rate. Lasers are useful in both gaining knowledge of structure and making the chemical reactions happen.

Chemical reactions are controlled by energy. Energy is necessary to make the chemical molecules of a substance react. For instance, a spark is required to make a mixture of gasoline and air explode to produce combustion in an engine. Molecules store energy as vibration; all molecules vibrate. The higher the vibration, the higher the energy level of the molecule. By steadily supplying vibrational energy to molecules, an organic dye laser can boost the reaction rate of some chemicals by a hundred-fold. This means that the chemicals involved will do what scientists want them to more easily and quickly.

ISOTOPE SEPARATION

Perhaps the most important application of laser-induced chemistry now being studied is *isotope separation*. An atom, modern physicists have found, is a complex packet of elementary particles considerably more difficult to explain than most of the simple models we use to represent it suggest. Essentially, though, an atom is built around a massive nucleus made up of protons and neutrons. Comparatively lightweight electrons circle the nucleus. Since the number of protons in the nucleus core of the atom always equals the number of electrons orbiting it, and the way an atom reacts with other atoms depends upon how many electrons it has, this proton/electron number gives atoms their atomic identity. Carbon—the building block of life, diamonds, coal, and oil—has six electrons and six protons.

In the early 1900s, however, scientists discovered that atoms with a set number of protons and electrons may have differing numbers of neutrons in the core. Stable carbon atoms, which we have mentioned, have an equal number (six each) of protons, neutrons and electrons. Carbon-14, however, has eight neutrons. This makes it intensely radioactive—a boon to science in one way. Carbon-14 dating helps scientists date the age of rocks, fossils and archeological artifacts by calculating how much the carbon-14 atoms have decayed (how many neutrons have been lost over time).

Atoms with equal numbers of protons but different numbers of neutrons are called *isotopes*. While the isotopes of an element are chemically close, they have nuclear properties that differ. The construction of the atomic bomb depended upon this difference.

The most abundant form of radioactive uranium, U^{238}, has 92 protons and 146 neutrons. The less common isotope, U^{235}, has 92 protons and 143 neutrons. U^{235} will fuel the chain reaction necessary for generating atomic power or the chain reaction of an atomic explosion; U^{238} will not.

Because organic dye lasers can be tuned to excite only the desired isotope of an element, they can sift various isotopes. This is called *isotope separation*. Current methods of extracting U^{235} from uranium compounds only get about half the isotopes available. Laser separation may get it all. This could double the fuel supply for nuclear energy obtained from uranium. For this reason, the process is called *uranium enrichment*.

In the future, scientists believe laser-induced chemistry may make it easier to produce rare drugs and chemicals needed in many fields. The long, careful preparation of many rare chemical compounds adds to their expense and lack of general use. In addition,

C.K.N. Patel adjusts gas flow into a high-powered gas laser at Bell Telephone Laboratories Murray Hill, N.J., Laboratory. The laser (which mixes helium, carbon dioxide, and nitrogen) has produced continuous outputs of more than 106 watts—the highest continuous output obtained from any laser to date. (Credit: Bell Labs)

lasers may enable power companies to salvage important radioactive chemical elements from nuclear wastes.

FUTURE SHOCK

Change occurs rapidly in any technological field once basic methods are outlined. Laser research is a booming field. Conferences on new uses of lasers in and out of the laboratory result in hundreds of scientific reports each year.

Bell scientist C.K.N. Patel sums it up nicely when he notes that "everything goes together. In order to accomplish something, one must know all about it, make advances in scientific understanding, in materials technology and so on. To go to shorter wavelengths (and make x-ray lasers) you must understand what sort of materials you're going to use, which means you have to understand the properties of the materials. What limits progress is simply the need for increasing knowledge about things you don't know." Lasers, with their unique ability to probe the unknown, are helping scientists blast away the limits.

Chapter 9

Light Show:
Lasers in Entertainment

THROUGH THE LASER CRYSTAL BALL:
LASER LIGHT EXTRAVAGANZA

As dusk falls, rich orchestral music vibrates the air with the lush sound of strings . . . and the lights begin to dance. Rising in intensity as darkness falls, the harmonious music of the orchestra is punctuated by the syncopated beat of a popular rock band, accompanied by flashing ribbons of laser light—green, red and blue streamers, intricate moving patterns exploding like fireworks.

Softening its role to a thrumming background accompaniment, the orchestra fades; the delicate, dancing ballet of lasers disappears, and amplified rock music pounds the air. The lasers sculpt light forms in the now dark sky. Then, the rock music stops; for an instant, there is silence, darkness.

Suddenly, to the brassy jazz trumpet, racing clarinet and throaty sax of the best jam musicians in the country, a bluesy voice is heard. But no vocalist is on stage. The audience looks up, their attention grabbed by a shimmering laser stage in the sky. There, an immense holographic image sways. The stage lights, themselves diffused lasers, dim. One by one, holo images of the musicians form in the night sky, until the entire orchestra appears to be performing on clouds. Then, magically taking shape from light like storybook nymphs, the dancers enter, and the audience witnesses the first sky ballet on a stage of quite real stars.

First produced in Los Angeles in 1973, laser light shows proved popular immediately. Numerous rock bands, several "Pops" orchestras, and planetariums in 20 cities throughout the U.S., Canada, Britain and Japan, enhance their shows with dancing laser beams. Even discotheques have replaced conventional strobe lights with lasers in some areas.

133

This is an image taken from one of the popular Laserium shows, created from laser light interacting with prisms, scanners and oscillators. (Photo courtesy of Laser Images, Inc.)

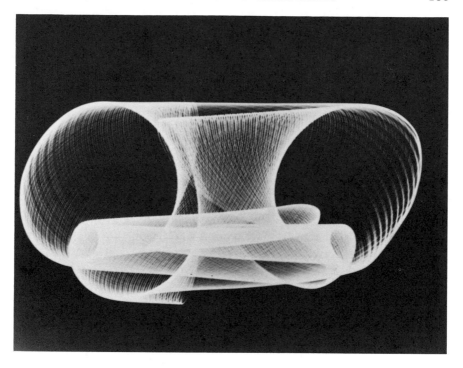

The images are controlled by a laserist who literally "plays" the light into these beautiful forms. (Photo courtesy of Laser Images, Inc.)

Reviewers have described the effects of these shows as "gauzy underwater dream shapes," "a celestial traffic jam," "clusters of jewel-like webs," and "a spiral-like Mobius tornado unwinding."

Currently, "laserists" create these brilliant, multicolored images which throb, dance and shapechange to the rhythm of rock bands or lush sounds of an orchestra, with argon and krypton gas lasers. These gas instruments, working something like fluorescent neon signs (only at laser brightness), provide exceptionally pure colors. Laserists change the shape of their beams with a variety of optical devices that diffuse, reflect and project the spears of ultra-pure light.

Prisms, mirrors, scanners and other components within the projector (which resembles a planetarium unit) control some effects. Bounce mirrors hung above the audience create other effects by making one beam appear to be many.

A special scanning projector can twist the light into intricate patterns. This unit splits the laser beam into horizontal and vertical lines that clash at right angles, resulting in swirling figures called *lissajous patterns*. In science, the occurrence of these figures on a cathode-ray oscilloscope verifies that two vibrations are of the same frequency. In a laser light show, the egg-shaped knotty lines twist in time to music, spiraling tails winding through vibrating figures.

WEBS, CLOUDS, MOSAICS

By beaming the laser's light through optical lenses of various kinds, webby diffusion images can also be created. Appearing three-dimensional because of the ultrabright clarity of laser light, diffused images range from a spread of cloudy luminescence to highly symmetrical mosaics. Each vibrant color is so sharp and bright it causes the eye to focus in such ways as to give the images depth and distance separation.

The effects at the laserists' control are varied enough now for John Tilp, who develops such shows for Laserium, to call them "sculpting with light." Laserium, responsible for what it calls "the original laser light show," first produced at Griffith Park Observatory in Los Angeles, pioneered the laser entertainment industry.

"The laser is the best tool we have to paint music with light," asserts Tilp, "but the idea of putting the two together has been around for a long time." In 1720, the Jesuit priest Louis Castel constructed what he called the Clavessin Oculaire. Played in a dark room, it admitted sunlight through prisms and colored, clear tapes to mix music and light. Rimski-Korsakov, Schubert and Beethoven assigned colors to musical keys, but lacked the technological means and expertise to match them. The Russian pianist and composer Alexander Scriabin created the little known piece "Prométhée Le Poème du Feu" in order to put the so-called color organ to use. At its first performance in a darkened Carnegie Hall in 1915, the organ projected colored light onto a screen as the music played. The word "lumia" was coined by Thomas Wilford in 1922, to describe his more freely associated use of color and music.

Writing in *Future Life* magazine, author Ned Madden foresees a future in which:

" . . . Major universities will offer courses in 'lumiagraphy.' Students will study the history of color in paintings, prints, and poems, and the art, myth and magic of the rainbow. They will study the physics of the spectrum, light, color, and the stars. They will study the human eye-brain, and the anatomy and psychology of how we see.

They will learn about the motions of the universe, harmony and harmonics, the music of the spheres, and the search for the spectral song. The common denominator will be the laser beam."

SAFETY FACTORS

Following a laser light show in Great Britain, the *London Observer* ran a cartoon depicting a group of punk rockers with guitars, giant amplifiers, and laser units mounted on their shoulders. "We deafen them, why shouldn't we blind 'em," reads the caption. The British science news magazine *New Scientist* editorialized that perhaps governments should "require a license to drive a laser."

Laser safety, however, is really no joking matter. Misused laser beams pose a serious threat to vision. "Contrary to the impression given by a popular James Bond movie, laser beams flashed about in a light show" will not saw anyone in half, Annabel Hecht points out in the *FDA Consumer.* "What can happen," she explains, "may be so subtle the victim may not even realize it has happened." If it enters the eye at the proper angle and intensity, a laser beam may burn the retina, which could cause a hard to notice impairment of peripheral vision. A direct hit on the eye's focusing lens, however, could create a blind spot that seriously impairs sight.

The Federal Drug Administration (FDA), which regulates radiation-emitting devices, classifies lasers in four categories. Class I lasers emit beams that have not been shown to produce any biological harm. Class II products emit beams strong enough to cause eye damage after long-term exposure. Class III lasers can damage human tissue with one short pulse if it strikes the eye directly. Class IV lasers are so powerful that even diffuse, indirect exposure to the beams might cause harm.

Depending on the class of the laser, various radiation safety features are required. These include protective shielding, emission indicators, safety interlocks, warning labels and beam control devices. But the FDA is no longer confident these safety features are adequate for use of entertainment lasers, which were not common when the standards were developed.

At the time safety requirements were instituted, most uses of lasers for entertainment or artistic purposes were confined to planetariums, classrooms or exhibits in which conditions were carefully controlled. Class I and II lasers, with their low risk, proved sufficient for these uses.

The need for highly visible beams, though, led laserists to using Class III and even Class IV lasers. FDA investigations have shown that some light shows using these lasers were conducted in what the agency considers possibly dangerous ways. One performer, for instance, sprayed laser images directly at an audience from a fiber-optic projector on his wrist. Investigators reported that at one rock concert where lasers were used, a stage hand lit a cigarette from a beam. In another show, beams bounced dangerously from mirrored walls and highly reflective balls hanging from the ceiling. Beams that strike a mirrored surface can reflect back at nearly full strength.

Although it is a small probability, a laser beam entering the eye at precisely the right angle could singe the optic nerve, causing total blindness. Only a fraction of a second is necessary for such damage to occur.

Aware of such dangers, manufacturers of lasers used in light shows have voluntarily developed additional safety standards, and the FDA issued new guidelines and continues investigations. The agency's two primary concerns are that some "people in charge of producing rock music shows know more about the music than about laser technology," and that they are using "general-purpose lasers that were not designed or manufactured for entertainment purposes."

The FDA advises those thinking about attending laser light shows to follow these guidelines:

1. Call the local health department to find out whether safety checks have been made. In some areas, enforcement of safety regulations is lax or non-existent.

2. Never look directly or through binoculars or cameras into any intense light source, laser or otherwise.

3. "If there is any doubt about its safety, it may be better not to attend the show."

TINKERBELL AND HELL FIRE: LASERS ON STAGE

Rock musicians, orchestra directors and amusement park operators are not the only people in the entertainment field who recognize the flash that lasers add to an act. Lasers have provided "performance lighting" and special effects for theaters from Broadway to Greensboro, North Carolina.

In a musical production of *Peter Pan* on Broadway, a computer-controlled, 10-watt argon laser beam played the role of Tinkerbell.

Floating through 15 flight patterns, the butterfly-shaped fairy created by the laser appeared suitably otherworldly. Six light frequencies, three blue-green and three ultraviolet, gave the image an eerie, translucent, fluorescent green hue.

Operator Randy Johnson tinkered constantly with the controls of his laser projector to keep the image in harmony with the live actors. His one-millimeter beam expanded to only the size of a half-dollar 100 feet from the projector, scanning the fluttering butterfly shape hundreds of times per second.

No less temperamental than other actors, the laser Tinkerbell occasionally suffered "from pressure problems in the ionizing tube," and was operated at only one watt to keep the scenery from smouldering. But, each time she saved Peter Pan by taking the poison intended for him, and he asked the audience to revive the fairy by clapping as her image faded away, thunderous applause shook the theater rafters.

A team of scientists and engineers at Laser Media, Inc., a professional special effects firm, created Tinkerbell. At university theaters throughout the country, experiments to create other light actors are being conducted by theater arts directors.

Lasers for rock concerts and Broadway shows are expensive, ranging in cost from the two million dollar touring bus of M.A.S.S. Laser of Canada (which sports four krypton gas lasers to produce green, blue, red and yellow beams, and an argon unit for blue-green), to $8,500 units manufactured by Spectra Physics. Projection and imaging systems to put the lasers to work entertaining audiences can add another $5,000 to $10,000 to the bill.

At the University of North Carolina–Greensboro and the University of Missouri–Columbia, theater departments have come up with lower cost means to laser projection lighting and special effects. These experiments relied on the use of low-cost, low-power (3.5 milliwatt) helium-neon lasers to create impressive, floating-in-air special effects. At the University of Missouri, this system was put to a performance test in a production of Hell Fire II, billed as a visual experience in mime, laser and sound. Four helium-neon lasers beamed special effects synchronized to the music of Holst's The Planets and other music onto a cyclorama screen. Directors selected laser projections to enhance the movement of the mimes as well as the music. Flash pots, conventional lighting and trap doors aided in other special effects. Total production costs: $75.00, although borrowing of equipment accounted for many savings.

Chapter 10

Through the Looking Glass: Lasers and Holography

Time: the future. Place: the White House, Washington, D.C.

A major crisis threatens to plunge the world into long-feared nuclear war. Sitting at his desk in the Oval Office, the President slumps a bit in his chair. He has not slept for days and knows that unless the Russians can be calmed, war could begin at any moment.

He orders his advisors from the office, then presses a button on his desk console, opening once again the direct hotline to Moscow. After brief preliminaries, the face of the Soviet Premier appears on the President's viewscreen. "We must talk . . . face to face," the President insists. The Premier agrees. Both make the appropriate technical arrangements. A short time later, the holographic image of the Premier appears in the President's office, while a hologram of the President appears in Moscow, deep within the Kremlin.

Each man can see the burden carried by the other. Their haggard faces mapped with lines of worry, the puffy circles under their red, sleepless eyes, speak more plainly than words. Though each man wears an earplug that transmits instantaneous translations of what the other says, they communicate as much through gesture and nuance as they do in actual conversation.

The meeting lasts hours. Eventually, however, the problem is worked out, the crisis averted, because without leaving their offices during troubled times, two men could talk, eye-to-eye.

Although meetings such as the one described above are still beyond the capability of holography, scientists believe they will be possible in the not-too-distant future. Holography is just beginning to move beyond the infant stage.

THE BEGINNING

Searching for a way to sharpen pictures of very tiny objects taken with electron microscopes, Dennis Gabor instead discovered the principle of holography in 1947. A scientist at London's College of Science and Technology, Gabor found that by using a "reference wave" in tandem with another light source, he could record and recreate a three-dimensional light wave pattern. Unfortunately, the reference wave worked best with a coherent light source. No suitable, intense coherent source of light existed at the time, nor would it until the invention of the laser twenty years later.

With the advent of the laser age, however, Gabor's principle of "wavefront reconstruction" led other researchers to develop modern holography. Gabor, who won the Nobel Prize in 1971 for his 1948 paper describing holography (a word he coined from the Greek *holo*, meaning "whole," and *graph*, for "message"), has noted:

"The [holography] principle was in hibernation until the advent of the laser for lack of powerful sources of coherent light, and by 1962 I had booked it as one of my many failures. Since that time, ingenious physicists and engineers have discovered a startling number of applications; they have realized almost everything except what I set out to do: see atoms! The explosive development of holography, which started in the early 1960s at the University of Michigan with the work of E.N. Leith and J. Upatnieks, was a great pleasure for me . . . and a great surprise."

Gabor, who died in 1979, lived to see his invention find many uses, from industry to entertainment, and was himself photographed (or holographed) in 1976. He attended the Museum of Holography in New York City shortly after it opened in 1974 and expressed delight at what he saw.

HOW IT WORKS

First we will take a look at the holography process as a whole; then we'll make a closer examination of each part of the process.

To make a hologram, a laser beam is split into two distinct beams. One of these, which is called the *reference beam,* is expanded and directed toward a photographic plate or film. The size of the beam is enlarged to completely cover the film. The second beam, called the *object beam,* is pointed at the subject of the picture. It is also expanded so that it lights the entire scene to be holographed. Reflected from the

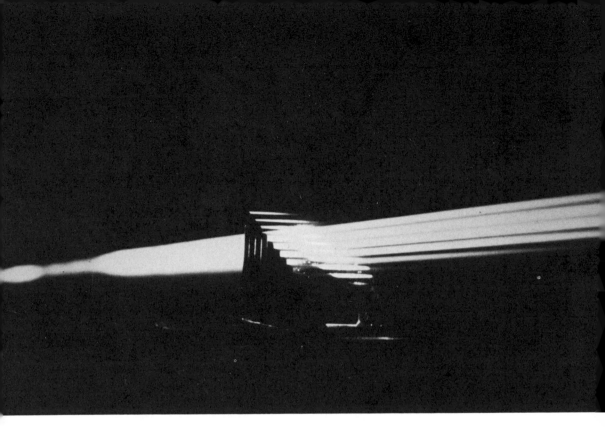

Various optics make it possible to achieve fancy effects with
a laser's coherent light. (Credit: Bell Labs)

subject, this beam is also directed toward the film, carrying with it the
visual information. The two beams meet at the photographic film,
where they create an *interference* pattern. This complex pattern is
what recorded on the film. When the film is developed, passing a laser
beam through it produces a three-dimensional image made of light.

The beam used to illuminate the hologram is essentially the same
as the reference beam used to make it. The light wave pattern from the
subject beam is very complicated; the light wave pattern of the reference beam is basically simple, particularly when coherent laser light
is used. Without the use of the reference beam, it would be impossible
to "read" the information placed on the photographic film.

Optics, such as lenses, mirrors and beam splitters, direct the two
beams. These must be properly placed. As you will probably recall,
light waves are, in one sense, tiny vibrations. Since the vibrations of
light waves are so small, holograms must be made in a nearly
vibration-free environment, so special tables and equipment are usually required. Otherwise, the image will be affected.

What is actually on exposed hologram film? When the object
beam from the laser bounces off of the composition, it showers the

film with light from all angles of the objects upon which it is focused. The reference beam, also striking the film, is blank. There is no information about the subject in this beam. Woven together as they meet at the film, the two beams *interfere*. This means the waves clash and merge. When wavelengths of the same size meet, they reinforce each other, creating a brighter image if they are *in phase* (the crests of one wave exactly matching the crests of another). A dimmer image is produced if waves meet crest to trough, for one reduces the effect of the other.

The interference pattern the two beams make when they meet at the plate is recorded on the film as something like a set of many small diffraction lenses. When a laser beam is passed through these lenses (the film) at the same angle as the reference beam orginially used, the light is reshaped by the lenses to recreate a three-dimensional image. If you view a laser-illuminated hologram, you will notice that the three-dimensional image can only be seen from certain angles. Move too far one way or the other and it disappears. These holograms must be viewed from the angle at which the reference beam struck the film and, later, lights the image. At present, this means that people at holography exhibitions are constantly bobbing up and down trying to see the images of this type of hologram at just the right angle.

APPLICATIONS

Although holography is still developing as a tool and an art form, it has already found many uses in entertainment, science, industry, art and education. Holography can capture and reconstruct the waves of sound and x-rays as well as those of visible light; this has made it effective as a scientific and industrial tool. Holography gives high-resolution (sharp) pictures of even very close objects, which is a problem with ordinary photography. Holography produces unrestricted natural vision of its subject without the need for special glasses or viewers.

ENTERTAINMENT

Holography caught the imaginations of many in the entertainment, art and advertising fields quickly. Many television shows and movies have used the *idea* of holography, if not, always, real holograms. Frequently, TV script writers for programs such as *Banacek, Switch,*

Six-Million-Dollar Man, The Hardy Boys and *Nancy Drew* mysteries have turned a plot on the idea of replacing a real person or object with a hologram. Fictional crooks often find this a helpful way to pull off a theft or provide themselves with an alibi. In reality, however, this might prove difficult at present levels of technology.

Holograms, recorded by monochromatic laser light, do not reproduce natural color or shade. Moving holograms are limited in scope—the viewer has to circle the image, or it must be mounted on a revolving device.

The three-dimensional image of Princess Leia that asks R2D2 for help in *Star Wars* was supposed to be a hologram, but actually was not. In *Logan's Run,* another popular science fiction film, six motion holograms actually were used when the computer of the movie interrogates Logan and his images simultaneously.

An amusement park haunted house at Ocean City, Maryland, projects floating holographic skulls near the ceiling. Others have used them to show witches holding crystal balls, ghosts that appear from nowhere to join riders and shimmering monsters almost too real.

Advertisers give away holographic jewelry and show holo images of their products at trade shows. One firm sells invitation cards (to trade shows) that incorporate a "Lucky Keyhole" hologram. A hologram of a man smoking a Salem cigarette amused travelers at New York City's Grand Central and Penn Stations in 1977. Ronald McDonald of hamburger fame is having a holo portrait done. Several large corporations have sponsored tests to evaluate further advertising use of holography, and we're sure to see an increasing use of it in the future.

INDUSTRY

Tire manufacturers can test their products without blowing them up through holography. Small changes in a holographic interference pattern can reveal separations, looseness and other problems. Acoustical holography (which uses sound waves) helps operators of tunnel-boring equipment to know what's ahead. This lets them work faster, yet more safely. Through interferometry, holography helps engineers test solid objects for stress. This is providing several industries with the means to check the quality of building materials, machine parts and so on. Holography is particularly useful here because it is nondestructive and all parts can be tested rather than just samples. By double-exposing holograms, engineers can see exactly

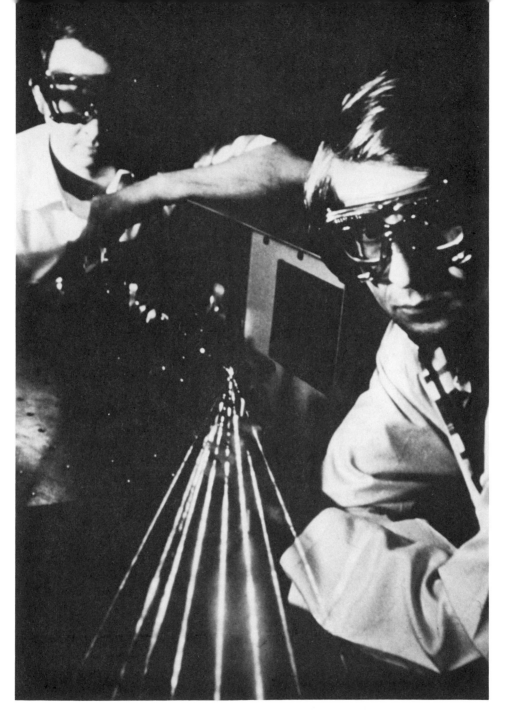

Some lasers will produce a rainbow of colors when their beams are channeled through optical devices. (Credit: Bell Labs)

where stress or deformation occurs, because the hologram is so exceedingly sensitive to movement that the slightest bending or stretching shows up. Thus, designers can make safer containers for storage and transport of nuclear materials, stronger tires or auto bodies.

One West German firm has developed a *Holocheque* I.D. card. Each includes a holo window with information (a fingerprint, photograph, signature, etc.), which would be very difficult to forge or duplicate. Another company in West Germany is studying a technique for using holography to see through fog, smoke or haze.

SCIENCE AND MILITARY

The Navy wants to combine holographic recording techniques and pulses of sound to get clearer pictures underwater. This could be useful for submarine navigation and hunting enemy subs. The Air Force is designing a holographic cockpit device that would allow jet pilots to see their instruments while looking out the canopy rather than down at the panel display. A NASA study predicts that 3-D images broadcast by satellites will enable future businessmen to meet without travel—much as suggested in the scenario that began this chapter. By transmitting a holographic image to a conference room, executives would save energy, time and money now spent on travel.

In medicine, researchers map the back of the eye with ultrasonic holography. This technique can also give doctors a view of broken bones and internal organs without the necessity of x-rays. Other researchers have used holography to view the interiors of crystals, and, at long last, see the arrangement of atoms. Although this isn't quite what he had in mind, inventor Dennis Gabor would have been pleased.

INFORMATION STORAGE

Storage of information is becoming one of society's major problems. The knowledge explosion has reached epic proportions. NASA alone receives 1,000 trillion bits of data from space satellites and probes each year. The smallest unit into which information can be divided, a bit is stored in binary language as either a 1 or a 0. Approximately eight bits make one byte—and a byte will represent one alphabet character (an "a" or "b" or "c," etc.), or two decimal digits (any two numbers in the decimal number system we normally use).

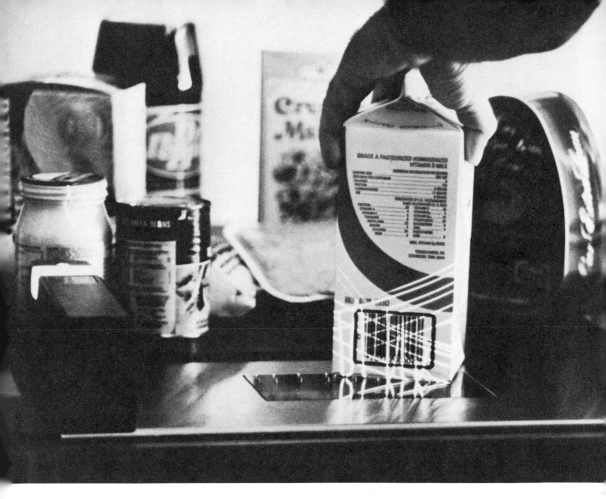

The new IBM 3687 scanner uses holography, a technique for creating three-dimensional images, to read data on packages such as the milk carton shown above. When a product that bears a Universal Product Code (UPC) marking nears the scanning window, a laser light "wraps around" the item. The laser beam's pattern is visible on the lower portion of the carton. These light patterns can "read" the UPC code located anywhere up to 180 degrees around the scanning window. Note, for instance, that here the UPC is on the side, not the bottom, of the milk carton. (Credit: IBM)

Industry today stores 265 billion financial documents, while the U.S. government alone stores 20 million cubic feet of records. New scientific information accumulates so rapidly each year that specialists are hard pressed to keep up in their own fields, let alone others. Since paper documents and books require so much storage space, we would all suffocate under a mound of the stuff—if we did not run out of trees first—if we tried to keep up with the ever increasing data flow that way. Computer punch cards, which can store about 800 bits of information each, are only a mild improvement, but not nearly enough. Magnetic storage of data on discs, spools, reels or cassettes is much better. A single reel of magnetic tape equals 677 boxes of punched cards. Tape, however, also has serious drawbacks. It has a short lifetime and is adversely affected by many environmental conditions.

Film, though, is sturdier. Microfiche film stores 88 pages of data—numerical, written or photographic—on a single four-by-six-inch sheet. It is so small, however, that dust, scratches and handling can wipe out important information. Therefore, many companies are now exploring the possibility of storing data on holograms.

Theoretically, a hologram only one square millimeter in size could store ten million bits of information, more than either magnetic tape or microfiche. But it offers other benefits as well. Since light from each point of the subject spreads across an entire hologram, the information is repeated. If the hologram were sliced in half, it would still be possible to retrieve the data. In fact, one would have to nearly obliterate the entire hologram before destroying the message completely. Three companies have already developed technology for storage of data on holos.

Classic movies rotting away in vaults are being saved by holo storage, which has the added plus of reproducing the colors of the original exactly. Some futurists predict that a combination of advanced computer technology, microminiaturization of electronics and holo storage of data may lead to elimination of ordinary libraries. At least one has suggested that physical schools, colleges and business offices may be gradually replaced by holographic recreation at home.

DOES THE MIND STORE HOLOGRAMS?

Modern physics, particularly the quantum mechanics often discussed in earlier chapters, has shown that light, matter and life all appear interconnected. All things, even those which we don't normally think of as objects, such as light, are in constant motion. The atoms of our bodies, this book, and the light you are reading it by are vibrating.

In a way, a hologram reveals this if you try to touch it. Your hand, of course, passes through the light image. What we see as round and real is actually light reflected from the object. We see an interaction of light and matter, not the object itself. If that sounds confusing, don't worry about it. Even modern physicists have a difficult time explaining such things clearly, because they have found there are definite limits to just how much we can know of reality.

The holographic process, however, does seem to mimic the way the human mind itself perceives reality. One scientist, Dr. Karl Pribram, a Stanford University neurophysiologist, believes the deep structure of the brain may be essentially holographic.

For one thing, the human brain seems to store information in the same way as a hologram: redundantly. Studies have shown that some of the memories, knowledge and skills in our brain are not, as once thought, stored in one specific spot. In addition, the holographic model of the brain Pribram suggests would account for "reality we perceive." We perceive the world, he says, through the "lens" of our brain, which imprints the information holographically on the cells.

Another scientist, Itzhak Bentov, goes even further. "We know that each of our cells can produce a human being with the same physical characteristics as the original," Bentov wrote in 1977, referring to cloning. "A holographic storage of information," he continued, would not only explain that, "but would imply that since we are a part of the universe, we also contain all the information contained in the universe itself." To know the universe, Bentov said, we need "only know ourselves well." It is difficult, considering such a theory, not to recall the 1960s slang expression: Far Out!

Chapter 11

To Make a Sun: Laser Fusion; Lasers in the Future

Time: halfway into the next century.

Inside a camel-backed building covered by steel plate that folds to the ground like a tortoise shell, a nuclear engineer presses a button. Responding to his touch, a bank of ultra-powerful lasers bombards a tiny glass pellet. It contains two forms of hydrogen removed from seawater—deuterium and tritium.

The combined laser beams, focused by an intricate arrangement of mirrors and lenses, strike the pellet with a blast that lasts but billionths of a second. Yet they deliver trillions of watts of power—more than all the power plants in the entire United States could generate in the same period of time during the energy hungry 1980s. Crushed to nuclear dust by the laser blast, the pellet falls in on itself with such force that it releases nuclear fusion energy—the energy that drives the furnaces within the sun and stars.

This fusion power, this tiny sun, supplies vast cities with light, warmth and a multitude of electronic conveniences. It makes cheap ethyl alcohol to run cars, trucks and buses—a good thing, too, for the world's oil reserves are nearly depleted.

But this night it has another job to do. Pulsed to an array of laser cannons surrounding the Enterprise III, this star power bursts from their muzzles as a dazzling spray of powerful laser light. Hitting a circle of mirrors around the spaceship that tilt toward the ship's base, the laser power builds. Then, a brilliant red pillar of focused laser light lifts the spaceship upward. It rises slowly at first, then ever more quickly, until, riding a laser torch, it disappears from the sky. Moving quickly enough to pull away from the earth's heavy gravity well, it shoots into space. There, the ship's captain barks orders, and soon a fusion flare, a controlled flash of starpower, boosts the ship toward Marsbase.

Model of 20-arm, 25-trillion-watt Shiva laser for fusion research at Lawrence Livermore Laboratory. (Lawrence Livermore Laboratory)

Once again, we must admit that this is science fiction . . . now. Yet laser-created fusion power, energy derived from the same source that fuels the sun and stars, can already be achieved. Both the United States and the Soviet Union have funded billion-dollar research programs aimed at producing inexpensive fusion power by the end of this century. Although estimates as to when this will be possible vary, experts agree that it *will* happen.

Even the laser-powered spaceship portrayed in the scenario above has been seriously proposed as a way to make space travel more economically feasible. One of the things that prevents man's expansion into the "final frontier" of space is the currently very high cost of launching rockets.

WHY FUSION?

The nuclear forces that hold atoms and their parts together are the strongest known. Releasing the energy of atomic forces through a fission reaction in an atomic bomb or the combined fission-fusion thermonuclear explosion of a hydrogen bomb demonstrates just how powerful it is. A boxful of material exploded in nuclear reaction could destroy a city. One cubic kilometer of water contains enough nuclear energy to exceed all known reserves of oil.

Fission splits atoms to release nuclear energy. Fusion melds atoms together to do the same thing. The secret of releasing fusion energy is to meld (fuse) two deuterium nuclei together, forming a helium nucleus. But to do this, extraordinary temperatures (more than 1,000,000 degrees centigrade) are necessary. Why, you might ask, does equipment not melt to the ground if these temperatures are needed? Answer: Because the amount of matter used is so small and the instant in which the lasers deliver their blast of heat-energy so swift; slightly more than a quart of deuterium has but 10,000 calories, hardly enough to boil water.

The biggest problem in obtaining fusion is not supplying the heat-energy, which lasers can do, but containing the hydrogen isotopes as they near fusion. As it gets close to fusion, the hydrogen becomes what physicists call a *plasma*, a gas with the electrons of its atoms stripped away.

If a plasma—which is the stuff of stars, including the sun and the bright tails of comets—touches any other object, it gives away its heat to whatever it touches. Fusion will not occur if the plasma gives up its energy before the nuclei of its naked atoms meld. The enormous pressures, density and heat at the center of stars are neither possible nor desirable for controlled fusion reactions. Creating fusion without them, however, requires some fancy engineering feats.

One way that plasma can be contained—kept from colliding with other objects—is by putting the glowing gas into a "magnetic bottle."

Normally, hydrogen gas is unaffected by magnetic fields. When heated to an incandescent plasma, though, its stripped nuclei are positively charged, while its free electrons (ions) are negatively charged. Since each has an unbalanced charge, they are affected by magnetic fields. Several methods have been developed to contain the plasmas with magnets long enough for laser energy to spark fusion implosion. The most promising is a Russian invention called the Tokamak (a Russian acronym for toroidal magnetic chamber).

The Tokamak, which is also being studied in the U.S., consists of a doughnut-shaped device (toroidal comes from the Latin *torus*, for a

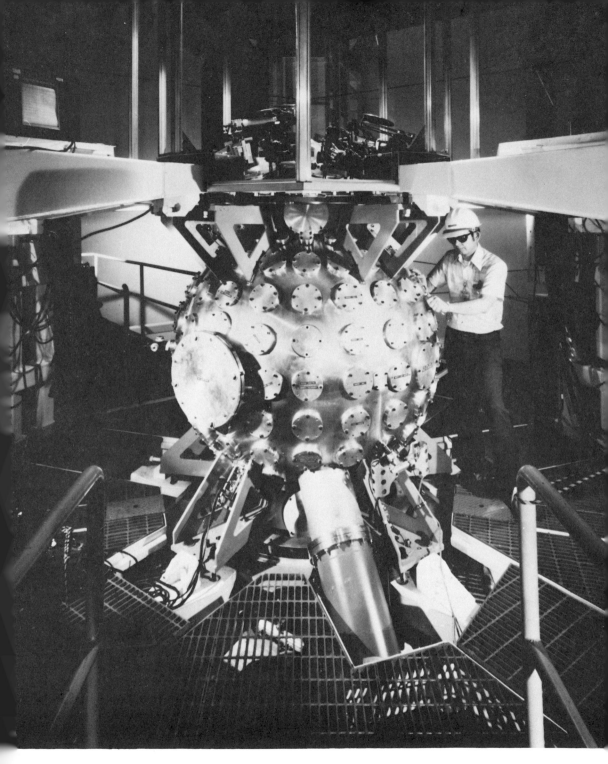

Shiva target chamber, Lawrence Livermore Laboratory. Ten beams come in from the top, ten from the bottom, simultaneously irradiating a hydrogen filled target positioned at the center. (Credit: Lawrence Livermore Laboratory)

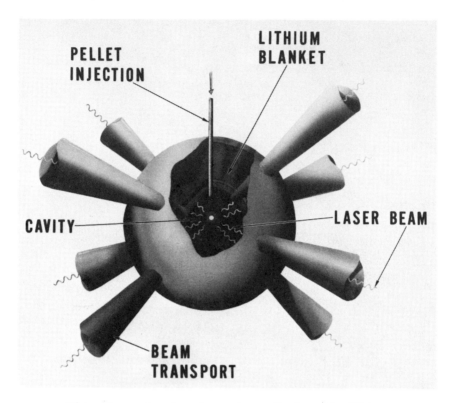

This cutaway drawing of a reactor cavity shows the lithium blanket surrounding and the area where laser beams will smash a hydrogen pellet. The plastic model shown here illustrates how a fusion reactor's one-kilojoule lasers will be positioned so that their beams focus on the same point, intercepting and imploding tiny pellets of heavy hydrogen to create heat energy. (Credit: Lawrence Livermore Laboratory)

round shape; it is now used to refer to doughnut-shaped objects in physics). The American version, set for operation at Princeton University by 1982, consists of 20 giant copper coils which conduct the magnetic field. Developed by the U.S. Department of Energy (DOE) at a cost of $284 million, it requires a cooling system that runs 8,000 gallons of water a minute through the coils.

Inertial fusion is a different road to star power. It relies on blasting the tiny fuel pellet with a huge battery of laser guns to raise its temperature so quickly that the fusion reaction can occur. In the

This is an overview of the Lawrence Livermore Laboratory's 2-trillion-watt Argus fusion research laser. A master oscillator—the L-shaped box on the center table—generates a ten-megawatt laser pulse lasting less than a billionth of a second. The pulse moves toward lower right and is split in two. At the corners, the two pulses are reflected along parallel paths toward the target chamber behind the wall at upper left. The equipment you see between the master oscillator and the target chamber as the pulses pass through are a series of amplifiers, filters, and devices to protect the laser equipment from being damaged by the extremely intense light and diagnostic devices to analyze the beam and its effects. (Credit: Lawrence Livermore Laboratory)

A 1,000-joule, neodymium glass laser chain for controlled thermonuclear fusion research is under development at the Lawrence Livermore Laboratory. In this photo, Dr. William Krupke, Assistant Laser Division Leader at Livermore, has his hand on the powerful flash lamps that "pump" the neodymium atoms to the energy states needed for laser action. (Credit: Lawrence Livermore Laboratory)

Soviet Union, scientists at Moscow's Lebedev Physical Institute designed devices that used a tangle of 216 lasers channeled through a maze of mirrors and lenses to the target.

In the U.S., scientists built several supermachines in attempts to produce inertial fusion (the name comes from the fact that when the target temperature is blasted upward very fast, the inertia of the atoms prevents them from expanding too quickly to stop fusion). The U-shaped, $3.5 billion Argus laser, a set of pipes lining a room the size of a high school gym, is one. It fires a neodymiumglass laser, split into two beams down the U-shaped tubes and amplified by a series of lenses to strike its target with four trillion watts of power in a billionth of a second. Next came the Shiva system. Named after the multi-armed Hindu God, Shiva, Lord of creation and destruction, it has 20 laser arms that deliver 10 times the power of the Argus unit. Shiva cost $25 million and is 360 feet long.

By 1985, Shiva will be incorporated into a still larger, more expensive, more powerful system called Nova. Nova will deliver a jolt of 120 to 300 million watts of power to its miniscule target. Designed to be 220 feet long, with a 100-foot target chamber, Nova needs 6-foot-thick concrete walls to protect workers from the searing heat of its laser blast.

What does it sound like to hear the lasers do their work in these giant machines? Disappointingly, perhaps, considering all of the other "bigness" they suggest, the lasers fire with a "pop" no louder than a firecracker. But the tiny flash of green light they create from fusing together atomic cores is a miniature sun.

The primary difficulty with fusion power at present is that the huge machines pump in much more energy to create fusion than is returned by the nuclear reaction. Scientists hope to turn that around by the end of the century. Although most scientific predictions regarding fusion have been very much on the mark so far, it is possible that researchers will get lucky and achieve controlled star power even earlier. "Fusion research is not technology limited, it is funding limited," says Dr. Robert L. Hirch, chairman of a U.S. advisory panel on fusion commissioned by Congress.

It should be noted that the Tokamak devices, which use radio-frequency energy pumping rather than laser heating of the hydrogen fuel, appear to be the most promising type of fusion reactor at present. Even these involve the use of lasers in another role. In an Oak Ridge · Tokamak experiment, lasers measure electron densities, giving scientists important information about what is happening inside the reactor. One of the latest developments in fusion technology, however, shows renewed promise for laser devices as well. Again, the Soviets pioneered the technique, which involves firing the laser cannons at a thin foil of hydrogen fuel rather than at a hollow pellet.

LASER AIRPLANES

Fusion power, with its promise of an energy-rich future, may be the most glamorous possibility for widespread applications of lasers in coming decades, but many others are almost as exciting.

The publication *Aeronautics and Astronautics* has suggested that laser powered airplanes may fly the world's skies if oil prices continue to rise. Solar-powered lasers beamed from satellites would zap receptors atop the planes. Changing the laser energy to heat for propulsion, the planes would fly on a light beam.

During takeoffs and landings, the planes would burn conventional fuel. About five miles up, they would switch to the laser system. They would carry reserve fuel for emergencies, enough to fly 600 miles. Safety features would make it impossible for the laser to operate except when striking the plane's receptor, so passengers would run no danger of being accidentally cooked. In addition, special shielding would protect passengers and crew while the laser is locked in place.

A laser-powered plane could save 8,300 gallons of fossil fuel on each transcontinental flight. Designed to carry 196 passengers each, the planes would fly special routes to avoid highly populated areas and other air traffic. Two satellites, a power station and a relay, would be necessary for each plane.

The laser-powered planes might save energy, but they would not be cheap. A 300-plane fleet requiring 300 power satellites and 400 relay satellites would cost developers billions of dollars. Advances in laser technology, optics and the space program are necessary before they would be feasible, according to industry sources.

The U.S. Federal Aviation Administration is looking at an unusual Soviet laser landing system for airports. Granted a U.S. patent, the laser system is called Glissada. It uses a network of ground lasers near airport runways to give pilots visual guidance to the runway. The FAA once rejected a similar idea because the agency feared the low-power lasers might threaten pilots' eyesight; now, though, they understand the lasers' limitations and want the system for its simplicity and ease of use.

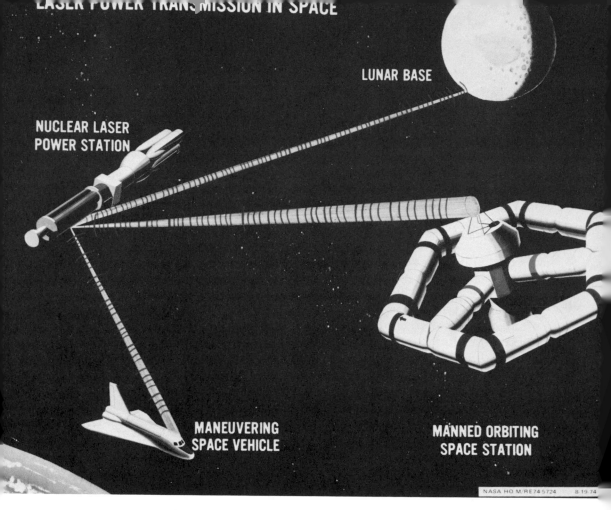

LUNAR BASE

NUCLEAR LASER
POWER STATION

MANEUVERING
SPACE VEHICLE

MANNED ORBITING
SPACE STATION

NASA HO M/RE 74 5724   8 19 74

A nuclear powered laser station, parked in a suitable Earth orbit, could beam power via lasers to various customers in space, to other space probes for propulsion, to lunar bases for their power needs, or even back to Earth, providing clean and abundant energy. (Credit: NASA)

LASER SPACESHIPS

The idea of laser-propelled spaceships is based on the fact that laser light can heat substances at considerable distances. Firing a ground laser at exhaust material—almost anything, even water—expelled by a rocket would rapidly turn the exhaust into a supersonic jet. This would provide enough power to boost the spaceship into near-earth orbit after it lifted several hundred meters off the launch pad.

In only a few minutes the laser-propelled ship would speed thousands of kilometers upward. Since escaping the earth's strong gravity well is what proves costly in weight and fuel in conventional rockets, laser propulsion might usher in the *real* space age. Once in orbit, spaceships could maneuver using inexpensive fuels.

Seriously proposed by space scientists, laser-driven ships might be smaller than the giant rockets and shuttlecraft used today, but many more of them could be launched. They offer the possibility of opening space to *people,* not just governments, the military and giant corporations. A single gigawatt (1,000 megawatts) laser could lift off payloads equal to those that 2,000 Saturn V rockets might put up in a year. Solar power, space colonies and scientific advances in space research may eventually result from laser pushed Conestoga wagons streaming into the "final frontier."

SPACE COMMUNICATIONS

One of the first U.S. Space Shuttle projects will be a test of a space-based laser communications system, called, appropriately enough, Lasercom. Lasers offer numerous advantages to space communications. Unlike radio frequency waves, laser light is almost jam proof, incredibly fast and difficult to intercept.

An Air Force project is planned to test a laser system that can transfer information at the rate of one billion bits per second. This, says a laser scientist on the project, is fast enough to transmit the contents of the Encyclopedia Britannica in one second. It is slightly less than the information capacity of 14 color-TV stations broadcasting at the same time. (Transmitting the color information in a TV signal uses quite a lot of "bit" space since the screen image is actually rapidly changing lines of dots).

Pumped by the sun, laser communications satellites could relay messages through other satellites to ground stations, submarines, aircraft and ships.

Far beyond this moderate use of space lasers for communication, in the future they may serve as links between ships traveling between planets, a moonbase and the asteroid belt. "It may well be that when the space age reaches maturity," Dr. Isaac Asimov wrote in "Let There Be A New Light," "a truly enormous load of information will be carried by laser beams interlacing space between the various human outposts." If this happens, he adds, future historians may claim that space exploration would never have "progressed beyond the primitive hit and miss stage without the laser."

Lasers have another very useful quality which is almost sure to be put to work in space. Since their coherent beams travel in absolutely straight lines and spread very little over long distances in the vacuum of space, they can be used to measure distance, spin, size and speed of other spaceships, asteroids and so on. Laser gyroscopes, too, would be more accurate than any conventional method of keeping space vehicles aligned for docking, landing and flight trajectories. While the laserlike weapons of *Star Wars* and *Star Trek* are certainly possible, it is likely that lasers will be performing less flashy but more important space duties as well.

THE EARTH MOVED

Many other space-related uses of lasers have been proposed by scientists at NASA, the military, and private corporations and foundations. A final mention—though certainly not the last you are likely to hear about in coming years—brings us back to earth. One of the space shuttle's experiments with lasers is designed to measure earth movement along the fault lines that cause quakes. Interestingly, all of the media attention space-laser weapons have received made NASA reluctant to call the device a laser. "We've taken to using euphemisms," a NASA official told *Sky* magazine. "People seem to think that with a laser in space we're going to be zapping everyone on the ground. So, we're calling it 'Space-Borne Geodynamic Ranging.'"

Already physicists from Maryland's Goddard Space Flight Center have demonstrated a precise system for measuring the shifting movement of fault lines with lasers. In California the scientists set up laser tracking stations on each side of the San Andreas Fault. Once every second the lasers pulsed light to a satellite orbiting 600 miles above. Bounced back to earth detectors, the lasers revealed that the fault is moving together faster than anyone thought. Tests in 1972, 1974 and 1976 showed that the two sides of the deep rift are closing in on each other at an alarming rate of three-and-a-half inches a year. Towns on nearly opposite ends of the state moved 14 inches closer in four years. In the future, such laser measurements may help predict when earthquakes are about to occur in time to save lives.

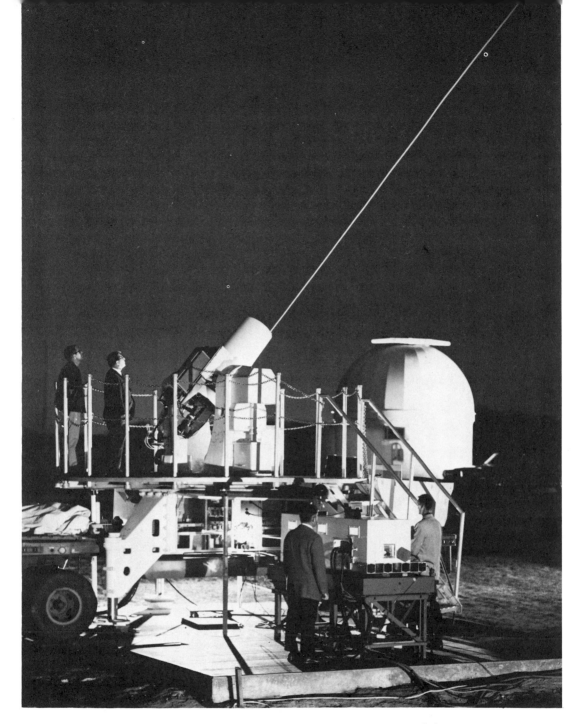

Engineers of the Goddard Space Flight Center, Greenbelt, Md., use this continuous argon laser to send a message to an orbited satellite. So far, they have been able to reach the Explorer XXXVI orbiting between 671 and 976 miles up. (Credit: NASA)

FINDING OIL SPILLS

U.S. and Canadian engineers tested a method for both finding and identifying the chemicals in ocean oil spills with lasers. An aircraft equipped with a laser beams its light onto a slick. The light causes the chemicals in the oil to fluoresce (light up), revealing the identifying spectrum of its elements in ultraviolet light. Collected by a telescope, the ultraviolet rays are analyzed by a spectrometer. This technique could help authorities determine what tanker caused the oil spill.

Another proposal by the same research team suggests that scanning the ocean surface with a laser by air could detect spills. If detectors sensed oil chemicals, an alarm would sound in the plane. Finding spills quickly via laser scanning would help scientists deal with them before they reached shorelines or seriously harmed sea life.

WHY WOOL MAKES YOU ITCH

New knowledge gained from lasers is revealing information almost daily about everything, from why wool is prickly to the links between genetics, the environment and cancer.

In Australia, where sheep are a major livestock and wool a major product, researchers used a laser analyzer to count the distribution of coarse fibers in wool. They found that the prickliness of wool is caused by a fraction of one percent of the coarse fibers present. With the aid of fast helium-neon laser analysis, textile manufacturers may be able to spin finer yarns and produce a warm woolen sweater that doesn't make you itch.

West German scientists have developed a holographic technique that helps plastic manufacturers design stronger materials and products. Tiny irregularities, such as cracks only a micrometer wide, are revealed in the disturbed interference pattern of a hologram. A design without weak spots shows a uniform, regular pattern.

With krypton lasers married to dye lasers, scientists are exploring how some cancer-causing substances change human DNA. Bathed in ultra-short pulses of the krypton laser's blue-green light, at least one common carcinogen found in smoke is being studied. "Things light up better with lasers—a lot of things," *Science News* reported not long ago. What happens is that laser light of the proper color makes many substances fluoresce. This occurs because atoms and molecules will absorb light of a specific color, then give off light of a different color that tells scientists what it is and where it is. Eventually, this may

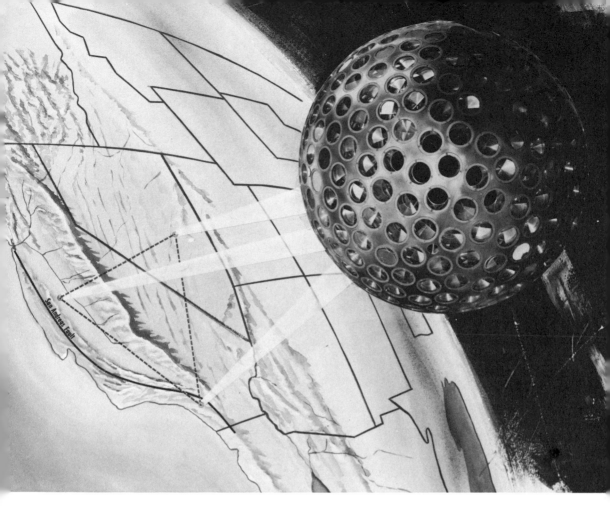

San Andreas Fault

NASA's LAGEOS satellite is helping to measure the north-south motion along California's San Andreas Fault, the juncture area where the Pacific and North American plates are grinding past each other. Laser beams reflected back to ground stations located on either side of the fault pinpoint the relative locations of the stations. Measurements taken over a period of months or years will show how the stations, and in turn the plates, are moving in relation to each other. The LAGEOS program could be the forerunner of a world-wide system for predicting earthquakes. (Credit: NASA)

make it possible for scientists to observe living cells in colors that reveal exactly what is happening chemically. Basic knowledge of this sort can lead to development of new drugs, cures for little understood diseases, ways to combat aging and more.

LIFE ITSELF

Combined with new knowledge gained from lasers, advanced laser tools for viewing and operating on individual cells may make yet another science fiction dream possible: genetic engineering. The wonders of the so-called "biological revolution" that would result could be the most amazing of all the laser's achievements. Cloning, the process of creating an exactly identical twin of any organism from a single cell, might be possible. Your son or daughter might be you.

Working on the stuff of life itself, lasers might blast junk out of your cells, clean arteries thick with deposits, "fix" cancer cells and help scientists vastly extend our life spans. Of course, the laser alone will not bring any of these things about; but it will play an important role in making them possible.

"The importance of lasers," says Gordon Gould, "is not in any one thing they do. It does things usefully in every kind of process. In being able to control and manipulate light, it is a universal instrument."

Dr. Arthur Schawlow, involved with laser science and technology for more than 20 years, probably understands the device as well as anyone. Still skeptical of their use as weapons, despite the billions spent in the U.S. and Soviet Union on "death rays," Schawlow cautions that they do, obviously, have limits. But in the 1979 Spring/ Summer issue of *Stanford Magazine*, he summed up their potential this way:

"As lasers have been improved, they have found an enormous range of important applications. As we learn more about how to make them and apply their subtle powers, they will inevitably find many more places in technology, and even in the arts.

"In ways that science fiction never dared imagine, lasers may serve us in the future. . . . Entirely new and radically different kinds of lasers will probably appear, and as our knowledge of light and matter grows, lasers will make practical what can barely be done today, and make possible what we have not yet even dreamed of."

Sources and Resources:
Finding out More

BOOKS

Dr. Isaac Asimov.   For more than 25 years, the good Doctor has tried singlehandedly to make science more understandable and enjoyable for the non-scientist. With over 200 books of science fact and science fiction to his credit, Dr. Asimov has written too many good books to list them all here. These are of special interest in the study of lasers.

Adding a Dimension. New York: Doubleday, 1964.

The sections on "Pre-fixing It Up" (Chapter 7) and "The Light Fantastic" (Chapter 10) help make sense of the metric system and the electromagnetic spectrum, respectively. Dr. Asimov explains the sizes of the wavelengths used in lasers, and their significance for future TV service.

Asimov's Biographical Encyclopedia of Science and Technology. Doubleday, 1964, revised 1974.

This invaluable reference work contains capsule biographies of scientists from ancient Greek times up to the present generation. Dr. Asimov has managed to compress the lives, ideas and discoveries of 1,195 men and women into an easy-to-read and fascinating book.

The Intelligent Man's Guide to the Physical Sciences. New York: Basic Books, originally published in 1960, revised and updated frequently.

Although this book doesn't discuss lasers specifically, it provides background information on many of the physical laws and phenomena that make lasers possible.

Is Anyone There? New York: Doubleday, 1956–1967.

In Chapter 11, "Let There Be a New Light," Dr. Asimov briefly discusses the history of lasers and speculates on some future applications.

Science Past–Science Future. New York: Doubleday, 1975.

View From a Height. New York: Doubleday, 1963.

"The Ultimate Split of the Second" (Chapter 9) explains ultra-tiny units of time, such as the picosecond.

Words of Science and the History Behind Them. Boston: Houghton-Mifflin, 1959.

Jeff Berner. *The Holography Book.* New York: Avon Books, 1980.

This recent book includes sections describing how holograms are made, the place holograms will have in industry and art, and comments from an army of front-running holographers. The bibliography contains complete listings of art galleries displaying holograms, schools offering courses in holography, laboratories producing holographic movies and companies supplying lasers and optical equipment. Recommended for students interested in holography.

Paul G. Hewitt. *Conceptual Physics: A New Introduction to Your Environment.* 4th edition. Boston: Little, Brown, 1981.

This bright, well-written physics text uses many illustrations and little math. Highly recommended.

Winston E. Kock. *Lasers and Holography, an Introduction to Coherent Optics.* New York: Doubleday, 1968.

Dr. Kock, a physicist who has been associated with laser research at Bell Labs, NASA and the Bendix Corporation, explains the physics needed to make lasers and holograms in this easy-to-read volume.

The Penguin Dictionary of Physics. Valerie H. Pitt, ed. New York: Penguin Books, 1977.

John R. Pierce. *Quantum Electronics: The Fundamentals of Transistors and Lasers.* New York: Doubleday, 1966.

Explains the advances in electronics that led to the maser, the laser and the semiconductor.

John F. Ready. *Industrial Applications of Lasers.* New York: Academic Press, 1978.

This comprehensive book takes a look at the use of lasers in many applications, including measurement, material processing, welding, drilling and cutting, holographic interferometry, spectroscopy, fiber optics, laser graphics and information storage. Assumes a basic knowledge of physics.

The Realm of Science, Volume 10: The New Science, Recent Advances in Physics. Stanley B. Brown, ed. Louisville, KY: Touchstone Publishing, 1972.

Chapter 8, an article on lasers written by David Park, focuses on the development of different types of lasers and their uses.

This electron-beam controlled carbon dioxide (CO_2) laser amplifier is part of the effort at Los Alamos Scientific Laboratory to control the same energy that fuels the sun and the stars—a thermonuclear reaction. (Credit: Los Alamos Scientific Lab)

The Techno/Peasant Survival Manual. New York: Bantam Books, 1980.

Written to "de-mystify" technology for non-scientists ("techno/peasants"), this readable and entertaining book explains the history and hardware of microprocessing, genetic engineering, the space program, weapons development, fusion energy, lasers and fiber optics. Read this book for a glimpse of the world of tomorrow being created by state-of-the-art technology today.

Alvin Toffler. *The Third Wave.* New York: Bantam Books, 1980.

Toffler speculates on the social, psychological and political implications of new technologies.

E.B. Uvarov and D.R. Chapman. *A Dictionary of Science.* 4th edition, revised by Alan Isaacs. New York: Penguin Books, 1971.

A guide to the many new words being created by modern technology.

Gary Zukav. *The Dancing Wu Li Masters: An Overview of the New Physics.* New York: William Morrow, 1979.

Without resorting to mathematics or scientific jargon, Zukav conducts us on a guided tour of the harmonious universe described by quantum physics and relativity theory. If you want to know why scientists get excited about their work, this is the book to read.

Index

WITHDRAWN